Developing Ecofeminist Theory

Also by Erika Cudworth

ENVIRONMENT AND SOCIETY

Developing Ecofeminist Theory

The Complexity of Difference

Erika Cudworth
University of East London, UK

palgrave
macmillan

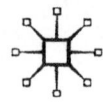

© Erika Cudworth 2005

All rights reserved. No reproduction, copy or transmission of this publication may be made without written permission.

No paragraph of this publication may be reproduced, copied or transmitted save with written permission or in accordance with the provisions of the Copyright, Designs and Patents Act 1988, or under the terms of any licence permitting limited copying issued by the Copyright Licensing Agency, 90 Tottenham Court Road, London W1T 4LP.

Any person who does any unauthorized act in relation to this publication may be liable to criminal prosecution and civil claims for damages.

The author has asserted her right to be identified as the author of this work in accordance with the Copyright, Designs and Patents Act 1988.

First published in 2005 by
PALGRAVE MACMILLAN
Houndmills, Basingstoke, Hampshire RG21 6XS and
175 Fifth Avenue, New York, N.Y. 10010
Companies and representatives throughout the world.

PALGRAVE MACMILLAN is the global academic imprint of the Palgrave Macmillan division of St. Martin's Press, LLC and of Palgrave Macmillan Ltd. Macmillan® is a registered trademark in the United States, United Kingdom and other countries. Palgrave is a registered trademark in the European Union and other countries.

ISBN-13: 978–1–4039–4115–2   hardback
ISBN-10: 1–4039–4115–7   hardback

This book is printed on paper suitable for recycling and made from fully managed and sustained forest sources. Logging, pulping and manufacturing processes are expected to conform to the environmental regulations of the country of origin.

A catalogue record for this book is available from the British Library.

Library of Congress Cataloging-in-Publication Data

Cudworth, Erika, 1966–
    Developing ecofeminist theory : the complexity of difference / Erika Cudworth.
        p. cm.
    Includes bibliographical references and index.
    ISBN 1–4039–4115–7 (cloth)
    1. Ecofeminism. 2. Feminism. I. Title.

HQ1194.C83 2005
305.42'01—dc22   2005047469

To the memory of my mother, Jill Cudworth (1935–2003), who taught me the importance of talking with canine companions, earning my own money, speaking my own mind and not worrying about "dust." Whose loving heart, acerbic wit and great hospitality are very much missed.

Contents

Acknowledgments		ix
1	**Introduction**	**1**
	Complex reality	2
	Difference and domination	5
	Gender and nature	9
	A map of the terrain	13
2	**Social Difference and Ecologism**	**16**
	Deep ecology and the negation of difference	17
	Social ecology: the problem of human domination	23
	Eco-socialism: capitalism and the commodification of nature	30
	Different worlds, different problems: environmental justice and liberation ecologies	34
3	**Complex Systems: "Nature," "Society" and "Human" Domination**	**42**
	Theorizing "nature"	45
	Nature, society and complex systems	53
	Anthroparchy: the social domination of nature	63
4	**Different Feminisms**	**71**
	Feminisms, systems theory and the problem of difference	72
	Deconstructing domination	81
	Patriarchy and the complexities of domination	86
5	**Ecofeminism and the Question of Difference**	**101**
	Women's affinity, men's distance? burning the straw women	104

Ecological embeddedness and the structuring
of social difference 114
Webs, spheres and systems: inter/relations
of domination 119

6 Embodiment, Materiality and Symbolic Regimes 128

Embodied theory 130
Alienated re/production 137
Enfleshed discourse: the naturing of women
and the feminization of nature 145
Contesting hu(man)ity: toward an enfleshed
posthumanism 148

7 Domination in a Lifeworld of Complexity 156

Complex lifeworld 157
Real dominations in social systems 165
Multiplicities of power: systemic permeability 170
The problem of difference revisited – life beyond
the matrix? 176

Bibliography 179

Index 213

Acknowledgments

The School of Social Sciences at the University of East London gave me a semester of sabbatical leave in 2004 to work on this book, and I would like to thank the colleagues who covered my responsibilities and those who approved my application. The artwork for the cover was produced by talented friends – postgraduate researcher and fiber-artist Rosy Collar, and photographer Tas Kyprianou. Their good-humored creativity and skill are much appreciated. Thanks to Dave and Dad for their love and support, and to Jake, for trying to be quiet, and not minding too much when I shut myself away to write – sorry. The non-humans of my household always prove good company. How anyone writes for lengthy periods without a cat or two on the desk (on the *paper*, on the desk …) and a dog underneath is a mystery to me. The 'girls' at work, Maria, Merl and Judith, engage with me in the important feminist praxis of laughing loudly at absurdity, gossiping and drinking wine. It always helps – cheers.

1
Introduction

> Is it possible, in the face of complexity, to construct theories, practices and selves that are mindful of connections?
> Chris J. Cuomo 1998:2

We live in an age where social and political theory is struggling to come to terms with both diverse changes and a multiplicity of difference. Ecofeminism is well placed to consider and confront a range of social inequalities and to theorize shifting formations of power. Ecofeminism can most simply be defined as a range of perspectives that consider the links between the social organization of gender and the ways in which societies are organized with respect to "nature." The most significant contribution of ecofeminism is the understanding of multiple kinds of social domination, of exclusion and inclusion based on varieties of hierarchies of difference (around class, "race" and place for example, in addition to gender) which both cross cut each other and enmesh, and which shape environment–society relations in important ways. The point of ecofeminist theory is to map the connections, the means by which formations and practices of difference and domination interlock. Such a project requires that we investigate possible points of unravelling of the tapestry of domination, and map the trajectories of strong and fast threads. The challenges to this project are many and varied, and this book, I hope, constitutes a strong defence of ecofeminism against its critics.

Weaving/spinning was a key metaphoric device of early ecofeminist literature (Daly 1979, Henderson 1983, Diamond and Orenstein 1990), and ecofeminism's many critics have focused on aspects of this work and may find the imagery irksome. Yet, the analogy is highly appropriate, as Mary Mellor puts it:

> A book on ecofeminism(s), feminism(s) and ecologism(s) must necessarily be a tangle of ideas, an interweaving of many threads. (1997:8)

Ecofeminism is not "only" about the relationships between the domination of women and of "nature." Given the huge parameter of their theoretical terrain, ecofeminists are terribly ambitious. The loom looms large and the threads are many. In being about what I call *multiplicities of domination*, that is, about intra-human and extra-human domination and the intricate patterns of such domination, ecofeminism could be about everything in critical, social and political theory. The mapping of such interrelations of processes and institutions of domination is difficult, but whilst the multiplicities of difference which ecofeminist theory compels us to account for, are extreme in their diversity, we can broaden our conceptual repertoire in order to produce more inclusive social and political theory. My intention here is to develop some conceptions drawn from a variety of perspectives, and to suggest a *multiple systems approach* to theorizing the complexity of life on this planet.

What has characterized both feminism and ecologism is transdisciplinarity. This has been both a strength in terms of the breadth of analysis, whilst also a source of misinterpretation and dispute. It is no longer sufficient for ecofeminism to be dismissed by simply being labeled with "the insult of 'essentialism' " (Salleh 1997:xi), and with the most naive and simplistic understandings attributed to various theorists. One of the aims of this book is to reclaim and reframe certain kinds of ecofeminist knowledge. This means to contextualize it in its historical, geographic and disciplinary location, and assesses the possibilities for reworking of certain concepts and theories with the insight of recent developments in social theory. David Byrne (1998:159) suggests that our academic future may be one in which "the notion of separate and distinct fields of science no longer has any validity as an intellectual position." Likewise, the social sciences. The harsh distinctions between what is science and what is not are becoming less certain. The study of environment–society relations can only benefit from such innovations, and they have begun within the trans-disciplinary understandings of complexity theory.

Complex reality

This particular struggle to capture social complexity brings together structural and systemic approaches with those of discourse and narrative and acknowledges the interpenetration of what might, if I am permitted some "old fashioned" terminology, be called material and ideological or symbolic levels of analysis. When I speak of the "ideological level" I refer to the symbolic representation of notions such as gender and nature. These notions are not unitary, but assume a wide variety of forms and are constituted with and often through multiple discourses of "race," age, ability, class and other marks of difference. *Symbolic regimes* are both constitutive of and reflective of materiality and I adopt a basically foucauldian position in examining both the distinctiveness and interpenetration of discourses of gender and nature. Those second wave European and American (eco)feminists who first drew attention to connections between the mistreatment of women and nature contended that there were common concepts, meanings and practices which embedded some women and some aspects of nature in the same discursive understandings. Within this writing, the conceptual boundaries around humanness, femaleness and nature are contested, and the discourses linking environmental degradation to the mistreatment of women are clearly seen as symbolic regimes constitutive of relationships of (systemic) power. In developing a complex account of systems and structures, I draw on the notion of discourse in considering how symbolic regimes of domination and contestation play themselves out in the lived experience of some people, animals and other "stuff." The material level is where dominations assume physical form, often embodied in specific institutions and their associated practices, and I endorse the term "*embodied materialism*" (Salleh 1997) in looking at the embedding of socio-economic practice in corporeality. Ecological impacts are often experienced most directly and pertinently as effects on human bodies, and ecofeminism acknowledges that our embededness within the "environment" is derived from our embodied position as human animals.

In considering multiplicities of gendered and natured domination as an interpellation of embodied materiality and regimes of representation, I draw both ontologically and epistemologically on *critical realism*. The concept of the real, which exists and is potentially independent of our knowledge about it, is important in developing an embodied materialism. The separation of "things" from different kinds of situated knowledges about them, is important for any attempt to better understand the intersections of social difference and domination, and indeed, to argue for the existence of such "things" in the first place. I conceptualize an environment that consists of entities or beings with objective properties, with characteristics independent of social processes and the limits of human understandings (Cudworth 2003:25, after Bhaskar 1978:22). Critical realism is fruitful for analyzing human embeddedness in the environment, because the "beingness" of persons, animals and plants is conditioned by their context. In allowing nature-independent properties and causal powers, critical realists have been accused of essentialism (Burningham and Cooper 1999:300). Such criticisms are misplaced. Rather, it is those who advocate a strong social constructionism wherein "beingness" derives exclusively from social action, and "nature" for example is a series of cultural constructions with various social functions (Mcnaghten and Urry 1998:44–62) that might be so accused. The "critical" in critical realism explicitly involves the analysis of the ways meaning is differently constituted and constructed across time, space and place without reducing "stuff" (as Cuomo 1998:29 puts it so perfectly) to social meaning alone. In addition, the ecologism in ecofeminism demands an epistemology and ontology that is able to account for beingness, agency and becoming, which is beyond both human control and human construction, and in which the coalescing of relationships which shape the natural and social world are seen to be "real" and as having real affects (Archer 1995).

To use the term *structure* is to invite criticism for seeing the world as some sort of filing system into which "stuff" (in social and political theory, overwhelmingly people or groups of people) is made to fit in a deterministic manner. However, the kind of structural and systemic notion I am seeking here is of dynamic patterns of relationships, which make and remake themselves, and which do not operate in a fixed and determined manner with a certain pattern of development or evolution. In looking for such a conception of system and structure, some critical realists have been attracted by complexity theory as a means of analyzing the historically dynamic, spatialized and specialized nature of complex systems. I draw on this work in arguing for a notion of multiple systems of social domination that are dynamic, interactive and ambiguous and at the same time characterized by complex structuring (Prigogine and Stengers 1984) and emergent properties and powers (Byrne 1998).

Whilst feminism has challenged the duality of gender, ecologism questioned the distinction between the social and natural, animal and human, and ecofeminist thinkers have problematized many of the binaries of Western dualist thinking, *complexity theory* challenges yet another key dichotomy: the separation of chaos and order (Hayles 1991). The chaos of life is the mass of components or stuff, which, to the mind of the complexity theorist, can self-organize/order and critically interact to form structures that exhibit systemic properties (Maturana 1980).

Fritjof Capra (1995) has drawn on such notions in describing the "web of life" in which structures and systems, which are often amalgams of social and natural phenomena, are incorporated. Whilst some ecofeminists have used web analogies (Plumwood 1993) they have not explicitly used systemic and structural analyses, nor dabbled with notions of chaos and complexity, as have other forms of ecologism. Part of the feminist difficulty with complexity theory may be a silence on questions of gender, or distaste for the intense abstraction of some applications in the social sciences. Yet as Mellor (1997:1) describes, the ecofeminist perspective is one that see the natural environment, including humanity "as an interconnected and interdependent whole." Complexity theory offers useful tools for theorizing such systemic relationships, and for both the social and the "natural" sciences, it opens up new ways of thinking about and deploying the concept of "system." In addition, the relations between different systems are an important element of complexity recastings of systems theory, and within this, issues of diversity can be directly addressed.

Difference and domination

Strands in feminist thought have often been defined and indeed policed, by an approach to difference. Socialist, radical, liberal, black and postmodern feminisms have all had different albeit shifting takes on the question of women's difference from men and from other women, struggling with the interces of class, caste, age, ability, sexuality and locality. Deep ecology has been particularly criticized for ignoring the impact of our intra-human differences on our relationships with the natural lifeworld, whereas social and socialist ecologies have sought to grapple with some of these differences. In mapping the implications of difference, diversity and domination for ecofeminist theory, and in considering the similarities and differences between various approaches, I have almost inevitably deployed a taxonomy of feminism and ecologism. Such constructs as "schools" of thought are problematic in that they simplify argument, homogenize groups of thinkers and "may do violences to the nuances of a writers thought" (Evans 1995:8). Such taxonomy is a mechanism to display the richness and heterogeneity of ecologism(s), feminism(s) and ecofeminism(s), and is hopefully broken down by the drawing together of different conceptual frameworks.

Equality of rights, opportunities and condition has been foundational for Western feminism for the last two centuries, but since the late 1980s there has been rather a preoccupation with the question of difference, and the extent to which its corollary, sameness, can apply to different social groups. The influence of poststructuralism and postmodern theory led some feminists to focus increasingly on mapping the differences between women (Barrett 1987) and an increasing stress on intersubjectivity, to the differences *within* women (De Lauretis 1986:14). For some influential theorists, all formations of oppressive relation are linked to the extent that gender cannot be separated out from other formations of social difference around race, class and sexuality (Spelman 1990). More strongly, categories of gender and sexuality have been castigated as inaccurate descriptors of the complexities of women's lived identities (Butler 1990). The "deconstruction" of women and gender into multiplicities of difference has led to a particular kind of critique of theories which attach any solidity, relative permanence and patterning as "essentialist." Ecofeminist approaches that have been seen to strongly articulate similarities between women and nature were particularly singled out for critique (Alcoff 1988:188). Feminists attracted to postmodern approaches have sometimes been labeled "difference feminists," for their preoccupation is the impact of difference on identity, and the fragmented female self (Benhabib 1992). I consider however, that all feminism(s) are preoccupied in various ways with difference between and within women, differences from men and within men, the impact of gender on differences of race and class and so on. Some ecofeminists have used both the terms difference and oppression to understand the complex structuring of power relations, and this book attempts to utilize an understanding of difference to develop a nuanced analysis of formations of domination.

The varieties of ecologism are probably even more preoccupied with notions of difference, and the questions raised by environmental ethics hinge particularly on the "process of differentiation" (Moore 1994:2). Deep ecologists have questioned the boundaries of species, the instability of contemporary biological and social understandings and taxonomies of difference, and have argued that this is related to a systemic privileging of the human species. Social, socialist and feminist ecologists have questioned the "difference" of nature, whilst insisting that this also takes account of the differences incurred in being human. The way in which I use *difference* is as a way of observing and unpicking the discursively constituted meanings of "stuff," which is in a condition of being different or unlike. I do not assume social differences are physically constituted, although corporeality may be an important feature of the constitution of difference. For example, the physical dissimilarity between myself, my son and Thumbelina the cat, is not the only materiality of difference between us, for small children and domestic cats are embedded in relations of economic and legal dependency with adults and humans. Children and cats are also socially and discursively constituted as a regime of ideas and beliefs that differ over time and space. Being different does not necessarily entail being implicated in systemic relations of domination, but my conception of domination is based on the difference of certain groups of people or entities from others.

My conception of *domination* involves three levels or degrees of dominatory formations and practices of power. These are marginalization, exploitation and oppression. These formations and practices are predicated on difference. Iris Marion Young has used the notion of "oppression" as an overarching concept to describe "systematic institutional processes" which prevent people developing their potential (1990:38), through exploitation, marginalization, powerlessness and violence. I prefer the term "domination" as a descriptor for systemic relations of power, and see marginalization, exploitation and oppression as different degrees and formations of dominatory power within a system. I define domination as not merely pertaining to intra-human relations and formations, and similarly to Young's use of oppression, I see it as limiting life chances, or as Cuomo (1998:77) more elegantly puts it, inhibiting the potential of an individual organism, group, micro or macro landscape, to "flourish." *Oppression* describes a harsh degree of relations of dominatory power. The application of the concept is species specific, and animal rights theorists and some socialist ecologists have helpfully distinguished between species that can be oppressed and those that cannot, for example, "meat" animals within factory farming systems of production might be oppressed (Benton 1993). *Exploitation* in a context of dominatory power relations based on difference refers to the use of something as a resource for the ends of the user, and this can apply broadly, to forests, soils and continents as well as groups of people and certain non-human animals. *Marginalization* is the making and conceptualizing of something as relatively insignificant, and although potentially less harsh in its immediate effects of power, may be particularly pernicious due to its potential invisibility (Frye 1983, Lukes 1974). The use of these different kinds of formation of social domination helps to capture the nature of domination as multiple, for it allows individuals and groups to experience occupation of dominatory power positions and identities at the same time as they occupy those of subordination.

The multiplicitious formations of intra-human domination are not fully explored here. I have focused on gendered and natured domination because I see ecofeminism as particularly fertile ground for developing more inclusive theorizations. Whilst ecofeminists have developed theoretical models using analogies of webs, notions of structured interrelations of power, and have explicitly made use of the term patriarchy, they have not used the currently controversial concept of system, nor "adequately explored the connections amongst the various forms and functions of oppressive system" (Cuomo 1998:6). I argue that a multiple systems approach is best placed to capture the complexity of social life. Patriarchy, capitalism, racism and what I call "anthroparchy" are forms of systemic relations, amongst others, which influence social life, and interrelate in complex and often ambiguous, chaotic and uncertain ways.

Within radical environmental social and political theory, or "ecologism" (Dobson 1991), there have been some attempts to consider human domination as a coherent system of power relations with particular effects. Deep ecology has tended to see modern Western societies as "anthropocentric" (human-centered), conceptualizing the environment as a series of resources for human use. Anthropocentrism is seen to explain and justify practices that harm the environment. Whilst deep ecology has been rather preoccupied with developing an adequate theorization of the ecological self, some theorists have drawn upon systems science and recently have used concepts of systemic webs. I develop the term *anthroparchy* to refer to a complex system of relations in which the non-human living environment (i.e. organic entities such as animals, plants, soils, seas and contexts for life such as rock and ice scapes) is dominated by human beings as a species. This systematic conception involves structures, sets of relations of power and domination, which operate to different degrees and have different forms, and are resultant from normative practice. I argue that different aspects of the environment are differentially dominated. Whilst virtually all aspects can be seen to be subject to human control, and many to human exploitation as resources for human use, some parts of the environment (itself a homogenizing term) can be considered "oppressed."

Much radical feminism, withstanding the critique from socialist and postmodern feminism and a range of other social and political theory has persisted in conceptualizing contemporary Western societies as patriarchal, hierarchically structured around the principle of male domination. Ecofeminism considers that there is a close relationship between the domination of women and the domination of the natural environment, and the approaches generated in the context of radical feminism, have tended to argue that the domination of nature is an aspect of a system of male domination. Feminist theory has not always accepted the idea of a system of oppressive relations implied by the idea of "patriarchy," and in the present academic environment, the use of the very word is a dangerous strategy likely to ensure one is not taken seriously. A range of feminist approaches have either rejected a structural analysis of gender relations, or subsumed a structural analysis within the systemic relations of class-based domination. I am generally sympathetic to radical feminist analyses that conceptualize gender relations as both systemic and structural, but I argue that a system of gendered domination is intertwined in a multiplicity of systemic dominations based on various forms of difference. *Patriarchy* can be defined as a system of social relations based on gender oppression in which women are dominated and oppressed by men (Walby 1990). The conception of patriarchy adopted here is of a dynamic system of social relations of power, composed of a number of structures that result from normative practice, which both adapt and alter over time, place and space and are based upon aspects of the system of oppression. Feminists who continue to deploy a theory of patriarchy tend to be of the view that this can enable discussion of dominatory practices and forms of oppression and of shifting patterns of marginalization and inclusion.

We cannot capture the nature of complexity with the notion of a single systemic formation of domination. Whilst the dominations of nature in its seemingly infinite variety, and of women in all their diversity are linked, I suggest this is the product of the relationship between interconnecting systems, rather than the product of one overarching system of domination or dualistic worldview. A multiple systems approach may be able to account for difference in both degree and form, of the oppressions based on gender and "nature." Systemic formations of domination have boundaries and can potentially be discrete, yet most usually, they interrelate and intersect each other in complex ways.

Gender and nature

The concept of *gender*, as distinct from sex, has a history in second wave Western feminism as a relatively (not entirely, Rubin 1974) undisputed concept, which refers to the social construction of biological difference, the social construction of the differences between men and women (Oakley 1972:158, Barrett 1980:13). Whilst feminist theorists may be divided as to the forms gender relations assume, the majority would accept the concept of gender as a social construction, although the content of that construction may differ historically and cross-culturally. It is no longer acceptable to consider gender in isolation from other formations, practices and identities of difference. The increasing feminist appreciation of women's difference is that we cannot consider the formation of gender without also considering what else women (and men) are, and what other forms of social relations constitute their experiences. More strongly, some feminists argue that both sex and gender should be seen as socially constructed (Haraway 1991). Some have contended that sex is as much of a construction as gender (Fuss 1989, Butler 1993) and others suggest less strongly that the boundaries between sex and gender are permeable (Assiter 1994).

Yet the problematic for ecofeminism is that we cannot secede to a postmodern definition of a sexed body as a cultural construct. Bodies are inevitably socially constructed. Social policies to improve sanitation and nutrition may for example, increase the height and longevity of embodied humans. The simple recognition of differential anatomy does not necessarily map on to a discrete binary of sex, certainly not of gender, through similar discourses of difference. However, humans are an embodied species with sexed bodies however diverse the identity, sexual practice and worldviews of those in the body as it were (Gatens 1996). All the critiques made of the sex/gender distinction do not actually invalidate it. Many adherents to the sex/gender distinction agree with much of the critique, for example that bodies are both socially and biologically constituted, that the body is an active and constructive agent, that gender cannot be manipulated at will, that binary oppositions are problematic per se and that one between the sexes may well be heterosexist. Alison Assiter (1994:125) has bravely argued for a non-fixed "essentialism," arguing for a minimally necessary set of biological features, the presence of which enables us to identify a body as female, and whilst these may be essential to such identification in a particular time and space, they are not fixed by nature. I would endorse such a critical retention of the distinction of female "sex" as necessary for an embodied ecofeminism.

There are multiplicities of difference of sex and gender in terms of discourse, identity and behavior, yet the multiplicity is not without social form and pattern. Catherine MacKinnon (1996) has argued that *women* are universal in their particularity, and that the specificity of time, space, place and "culture" still enables some congruity to the category "woman" and the use of the concept gender. Rosi Braidotti (1991:131-2) suggests women are a collective singularity, and firmly resists the notion that the body is an effect of discourse, a purely cultural artifact severed from biology. Braidotti's picture of sexually embodied women (and men), whose situation in the body is constituted through multiple discourses of gender, sexuality and other forms of difference is useful for any discussion of both difference and domination within systemic relations of power. Even qualified use of the word "women" however, has been seen as problematically exclusionary (see Spivak 1987). Susan Griffin (1997) usefully suggests that the use of a word per se does not inevitably mean a narrow understanding or usage. The use of the word "woman" by feminists entails the "recognition of many varieties of women within the word" (1997:215). The problems of words as descriptors are not "erased by erasing the word" (1997:217), rather one reveals the limits of categories by their use. For Griffin, an ecofeminist understanding of meaning is one that stresses the context of meaning, and with this in mind, we can use gender and nature as descriptors of power, whilst both deconstructing and contesting the categories.

Recognizing the enormous diversity of experience and identity, we need to hang on to the concept of gender (and other categorizations of social difference) and not collapse them into an amorphous, and I think, liberal individualist, notion of multiple identities. Gender, and women are more than fictive constructions. Using the critical realist notion of "causal powers" attributed to social phenomena, we can have women, and difference, and gender. As Caroline New puts it:

> Women are female subjects, that is, social beings who live within and through the local meanings attributed to femaleness. The existence of people with some attributes of either sex, or who are sexually different, does not threaten the concept of femaleness, though it may challenge our dichotomous simplicities. Femaleness is a relatively enduring set of attributes, although it is still context dependent. ... A woman is not just a female, though, but a female *person*, and poststructuralists are right to say that "woman" does not always mean the same thing (and this is) true of almost every category in social science. (1998:364, original emphasis)

Gender, like other formations of difference, is both fluid and relational, and this is imperative politically and ontologically, for ecofeminism – an understanding of the material and symbolic intersections with other forms of difference. The interesting point ecofeminism raises for other feminisms, is that difference is not only human.

Whilst feminists have adopted "gender" as a category of analysis because it assumes that relations between men and women are socially produced and structured, some sociologists have tended to use the term "nature" to capture the same process with respect to society–environment relations (Mcnaghten and Urry 1995, Eder 1996). Whilst nature can of course refer to biology (a behavior "natural" to a species, such as roosting for hens), it is also a social construct. "Nature" refers to accepted standards of behavior for different "kinds or types." This term is often misused and biology conflated with culture in order to justify the domination of certain groups of humans and other animals, which have been designated "natural." Whilst accepting this common conflation, and the interpenetration of the symbolic and material, I use environment and nature to refer to different things. The *environment* refers to particular and multi-variant physical phenomena, which have an existence not only within but also outside, human imagination and knowledge. *Nature* refers to the differential symbolization and material institutions and processes that shape the lived experience of humans as distinct, for example, from other animals. In using these different terms, I hope to capture a sense of both the socially constructed difference(s) between the human species and nature, and the embedding of human communities within the environment.

These definitions are a product of Western paradigms and sensibilities. As Kate Soper (1995) has suggested, the wildness of nature is a key element in distinguishing it from the human world, a means by which nature is seen to have some degree of authenticity (Goodin 1992). "Naturalness" implies some degree of independence from human influence, yet if we take as global warming as given, there is no place that is not influenced by human societies. Categories of nature and environment are slippery inventions at best, but I do think it worthwhile to separate the non-human from the "built" environment, and the social, and to consider the ways in which ideologies and fantasies of nature map on to the varieties of non-human life. Diana Fuss (1989:6) has suggested that a conception of nature as a fluid, historicized, social construct, may be a way out of the (false) essentialism/antiessentialism debate, but whilst this may be so, we also need some conceptual acknowledgment that the non-human lifeworld is something more than this, and can be beyond the scope of our knowing and our imagination.

The environment is sometimes defined as living nature, the huge variety of animate non-human entities. For ecological scientists, it is also the context, the conditions for the existence of animate species in all their incredible diversity (Cudworth 2003). Whilst "environment" is inevitably homogenizing, it captures what ecological scientists see as the interdependency and reciprocity that binds animal life, plants, water, air and earth. Part of the definition of the environment involves the diversity of life and natural form or biodiversity (Wilson 1988). Whilst we need this terminology, it is tricky to properly capture the conceptual distinction that would help us to grasp different layers of meaning and forms of embodied practice. Ecofeminism and the environmental justice movement have been particularly challenging in this respect, to our notions of environment. Chemical bleach in nappies, the marketing of powdered milk in some of the poorest parts of the globe, toxic waste dumping and even employment training have become defined as within the parameters of environmental issues (Taylor 1997:50–3). Anthropologists such as Tim Ingold (1995, 2002) have demonstrated that the human species is ecologically embedded – our dwelling, living and working are predicated on our embedded and embodied situation. In seeing our condition as embedded and embodied, social theory can move beyond human exceptionalism (Catton and Dunlap 1980:34) and reconsider its anthropocentrism.

Therefore, I use the categories gender and nature, as distinct from women and environment. My usage is predicated on the assumption that these are overlapping and imprecise descriptors. Gender and nature are not concepts that allow us to capture necessary truth(s) about beings, objects or practices, but they are concepts with sufficient descriptive regularity and definitive power to contribute to the patterning of material formations and symbolic regimes of power and domination. The discursive networks of gender and nature are not unified and self-evident, but complex and dynamic, and can be understood in a careful and specific way. For all ecofeminist theorists, theories and conceptual frameworks concentrate on the relationships of power between different social and natural formations, they do not defend a notion of systemic and structured power. The structures and systems of which complexity theorists speak, are patterns of relations, and what this book proposes is a way of thinking about those relationships in a way that marries the complexity of all our difference with an understanding of relationships of dominatory power.

A map of the terrain

I begin with a process of separating out – looking at the "problem of difference" for ecologisms and feminisms. First, consideration is given to the ways ecologisms have sought to account for the enormous multivariate complexity of social structure, regional variation in human culture and the ways human formations map onto their non-human lifeworld, their animate and inanimate environment. Second, the ways in which feminisms, of hues other than green, have considered intra-human

14 Developing Ecofeminist Theory

difference and diversity is examined, along with an account of the fear of much feminism in considering the enormous range of difference implied by considering the non-human. Third, I provide a critical overview of ecofeminst theory, and consider the ways in which it has drawn on a range of feminist, ecologist, socialist and other conceptual frameworks in analyzing the similarities and differences, linkages and divergences, between different formations of domination.

Chapter 2 critiques deep, social and socialist ecologies with respect to their analysis of "difference," and considers the development of liberation ecologies in relation to the established ecologisms of the first world, whilst also suggesting the importance of a systemic theorization. Chapter 3 develops these arguments further by examining the work on systems thinking emerging with both scientific and social scientific analyses of "complexity." I suggest how recent scientific models in systems theory have developed useful conceptions that may be deployed both in the understanding of social systems per se, and in the analysis of systemic relations between human communities, other species and environmental contexts. I argue for a concept of "anthroparchy," a social system of domination of the environment, whilst concluding that we require approaches that may account for the interaction of a number of forms of social domination and exclusion in the analysis of environment–society relations.

Chapter 4 engages with feminist debates around the "problem" of difference. The chapter examines the question of social difference in socialist and black feminisms, the attempts of "dual" and "triple" systems approaches to capture gender, race and class, and the turn to postmodernism in the analysis of difference. It takes account of the challenge of essentialism levied against radical feminisms, and argues for the continued efficacy of a theory of patriarchy within a complexity framework. Drawing on some radical and socialist feminism, I suggest that patriarchy can be theorized as a system of gender domination, which is crosscut and overlapping with other systems of social domination. This critiques post structural and postmodern feminist analyses, whilst also suggesting the possibility of combining a discursive approach to gender relations within a broadly structural framework. A particular theme of the chapter is a notion of feminism as "nature blind." For much feminist theory, the non-human is a "Pandora's box" best left alone, yet it must be opened, and the complexity of the ultimate form of difference is a challenge which feminism must take.

Chapter 5 provides an analysis of ecofeminism, exploring the differences between postmodern influenced perspectives and the radical feminist influence on certain contemporary ecofeminisms. It argues that despite certain difficulties, some ecofeminist theory can be seen as "complex modernist" and "multiple systems" theory. I suggest that, with further development, such analyses may be most satisfactory in getting to grips with the complexity of difference. Chapter 6 synthesizes elements of green social theory with eco-feminist approaches, in looking at the question of material production of agricultural and industrial consumer goods and services and the reproduction of "life." It argues for a social theory that can account for biological embodiment and the embeddedness of biology in social relations. It also attends to contemporary representations of gender and nature and considers how some eco-feminist thinkers have analyzed cultural formations of gender and nature, and in particular, the permeability of symbolic distinctions between nature and the human. Popular conceptions of hybridity are key to understanding contemporary developments in material relations between humans and "nature." Any attempt to theorize our current condition as "posthuman" however, must be cognizant both of the multiplicities of difference in domination, and of the embodied and embedded constitution of social life.

The concluding chapter suggests elements of a multiple systems theory and demonstrates its potential usefulness though an elaboration of the theoretical connections between the different formations of social difference and human dominations over nature discussed here. I draw together the arguments for an ecofeminist approach that is at once structural, systemic and cognizant of complex difference between humans and formations of social nature. This sketch is not prescriptive but suggestive; as such it must be, given the richness, diversity, complexity and downright messiness of this planet we inhabit.

2
Social Difference and Ecologism

Green social and political theory, or "ecologism" has imploded the boundaries of social science by insisting that we consider the huge diversity of non-human life when thinking about "difference." Some ecologisms have raised thorny questions of the relationship between the multiplicity of human differences and the enormity of differences within, between and across other species. This chapter provides a critical consideration of four different kinds of ecologism. It examines how difference is theorized, and whether the treatment of difference is linked to social and/or human domination in a systemic way. I am particularly interested in the way theorists understand the impact of social difference and intra-human domination on formations of environment–society relations. This prepares the path for Chapter 3, which draws on developments in complexity theory in order to argue that our social relationships toward nature in modernity are characterized by systemic domination, and that the domination of the environment is crosscut by intra-human dominations based on difference.

It is useful to distinguish ecologism from "environmentalism," which is usually defined by its concern with conservation through technocratic means (Benton 1994:31). Bill Devall (1990:12) refers to "reform environmentalism," and specifies this by its assumption that existing social practices such as those associated with consumer capitalism can be modified in order to preserve the environment (Elkington and Burke 1987). For Carolyn Merchant (1992:1) it is a sense of environment–society relations as in some sort of crisis that defines ecologism and as such, Andrew Dobson (1990:13–18) is right to suggest that ecologism is, by definition, radical. Ecologisms have various conceptions of the human domination of nature, and consider that current forms of social and economic organization constitute a threat to the well-being of both people and planet. However, variants of ecological thought have different analyses of exactly how and why humans can be said to dominate nature, and of how social differences affect environment–society relations. There are sufficient differences to distinguish deep ecology, social ecology, eco-socialism and liberation ecologism, with respect to how they conceptualize the differences incurred in being human, and the ways in which these are manifest in, or tangential to, relationships between "society" and the environment. Ecofeminism makes its presence felt only in the role of critique here, and is elaborated in detail in Chapter 5.

Deep ecology and the negation of difference

Deep ecology suggests that human exploitation of the natural lifeworld is currently precipitating a crisis in environment–society relations. We have damaged the environment to such a catastrophic extent that drastic measures are necessary to halt such destruction, and this requires that we change the way we conceptualize nature. For Arne Naess (1985:256) the use of the term "deep" for this kind of ecologism is suggestive of an articulation of fundamental and normative presuppositions. These priorities involve the development of an "ecosophy" or "earth wisdom" which, in the context of late Western modernity, means a different way of seeing the earth in which humanity is understood to be intimately dependent on the natural environment (Naess 1973:103).

Deep ecologists argue that we must move from human-centeredness or "anthropocentrism" as the key structuring principle of social organization, to a nature-orientated "biocentric or ecocentric" (Naess 1990:135) way of thinking. The theorization of exploitative relations between humans and the environment as systemic, as anthropocentric, is an important strength of deep ecologism. We need an account of human domination which enables consideration of the ways forms and practices of human domination of the environment are both enmeshed in relations of human difference and power, and might, in particular instances, also be relatively independent of the effects of such relations. In elaborating the concept of anthropocentrism, and in particular, in seeking to develop a non-anthropocentric ethics however, deep ecology fails to account for difference and diversity both between and amongst human beings, and the non-human animate environment. This said, many critical responses to such reductionism have themselves been problematic. Deep ecology has been equated with "wilderness reverence" in the American context, and with the activist writings of controversial anti-intellectuals. If we examine more sophisticated articulations of deep ecology however, there is room to move on from a reductionist critique, and account for difference by combining an understanding of anthropocentrism or some such systemic formation of human domination, with the complex cross cutting relationships of intra-human difference and domination.

Anthropocentrism: the human centering of the social

Deep ecology adopts a systemic approach to understanding the organization and patterning of both social and "natural" life. From scientific ecology, deep ecology adopts the view that all processes are connected and human intervention in natural ecosystems cannot be without impact. Deep ecology also adopts a philosophically holist position, which theorists such as Naess derive from Spinozist ethics. Spinoza suggested that "all" of nature, including inanimate objects such as rocks, should be seen as part of the cosmos, which collectively work to maintain the integrity of the whole (Palmer 2001:49). Objects in what we would now call natural ecosystems are causally active and shape their environment, arguments that recall the systemic notion of James Lovelock's Gaia hypothesis (see Chapter 3). For Naess (1973), living beings of all kinds are "knots" in a biospherical net or field of "intrinsic relations." In this worldview, the net or web of relations is of paramount importance and the "system" as a whole cannot be reduced to its constitutive elements for as Robyn Eckersley puts it:

> the world is an intrinsically dynamic, interconnected web of relations in which there are no absolute discrete entities and no absolute dividing lines between the living and the nonliving, the animate and inanimate, or the human and the non-human. (1992:49)

Such nets or webs of relationships are incredibly complex and need to be conceptualized in terms of vast systems. In addition to the webs of relationships in which we are interlocked with a diversity of species and scapes, deep ecologists understand systemically that human society is structured in particular formations of relations with the "natural world." That system of relationships has been termed "anthropocentrism."

Deep ecologists tend to argue that contemporary Western society is anthropocentric, or human-centered in its organization and has a dominant worldview in which the non-human world is both conceptualized and treated in terms of means to human ends. What is problematized in particular is the enlightenment project of placing human beings and their faculties (especially reason) in a pre-eminent position with respect to the "natural" world. For deep ecologists non-human life on earth has intrinsic value (inherent worth), yet anthropocentric society is not cognizant of this and therefore excessively manipulates, exploits and interferes with the natural world (Devall 1990:14–15).

Anthropocentrism is usually considered to be the most deep-seated form of domination. Warwick Fox (1989) for example, asserts that human domination of the environment accounts for other forms of social domination. It is a priori, and if humans can abandon this most deep-seated formation, then other dominations will consequently be eradicated. Such an understanding cannot account for possible relations between forms of intra-human domination and the domination of nature, nor can it specify how forms shift over time and space. To suggest that the social domination of nature requires separate conceptualization from the differing forms of social domination that structure human society, is an important contribution. However, in analyzing that anthropocentrism in isolation from intra-human dominations such as those based on class, "race" and gender, deep ecologism moves toward an anti-humanist position where humans as a species are seen to be collectively responsible for environmental destruction (see Foreman in Chase 1991). We need this conception of a society structured around human needs and interests, and in which the environment is dominated, exploited, and in some cases, oppressed. Yet we cannot set up such a conception in the way deep ecologists currently suggest, for humanity, given all its differences of power, wealth, abundance and consumption, is embedded with/in environments with different relationships and impacts.

Intrinsic value and the ecological self

The denial of the differences within "humanity" becomes most apparent when deep ecologists consider the ethics of environmental concern. The most common basis for an environmental ethics is an argument for "intrinsic value," according to which natural objects and species are seen to have value in themselves rather than having value in terms of their functions for other things (Bunyard and Morgan-Grenville 1987:281, Fox 1986:7). In articulating what intrinsic value might look like, deep ecologists have tended to homogenize the diversity of the non-human animate environment and argue that "all" the environment has the same value and should be treated similarly.

For Devall (1990:70), human-centered environmental preservation ethics and nature-centered preservation ethics are not opposing positions, for all forms of nature conservation can be seen as human "self defense." Naess considers that "every living being has intrinsic value"

(1990:135), and he argues for a biocentric egalitarianism wherein all living beings, including humans, have equal intrinsic worth because all species are equal in their importance to the planet as a whole. Such an ethic of intrinsic value covers "all life," the whole environment has value in itself (Naess 1984:20). In order to socially embed such an ethics, some form of "ecological consciousnesses" (Bunyard and Morgan-Grenville 1987:282) which extends the notion of selfhood beyond individual identity is often proposed. An "extended" sense of self is derived from our experience of engaging with the natural world around us, and once it is internalized, we will be drawn into a sympathetic identification with the good of the whole (Naess 1989:20,174–5, Eckersley 1992:172). However, even humans with an expanded sense of self may see the survival of their own self-hood as commensurable with a degree of environmental exploitation (Dobson 1992:59), and of course, so individualized is the project of finding our ecological selves, that our theoretical and political imperatives will be different and possibly divergent. Whilst deep ecologists claim to desire the "widest possible" identification with nature (Devall 1990:35), in practice, they are advocating identification through the experience of their living space, or "home" bioregion. This localized sense of self might broaden our species-specific notion of community, and destabilize its humanist boundaries, perhaps preventing our identification with what Midgely (1996) calls "exclusive humanism." However, it does not imply identification with human and other communities across the globe and it does not address key questions of global difference, particularly with reference to the grossly uneven distribution of wealth and the well-being of both human and non-human populations. For Mary Mellor (1992:85), such biocentric egalitarianism is almost inevitably anti-humanist. Indeed, in reading the deep ecological literature, human beings in all their dynamic diversity and dramatic difference, rather disappear from view.

Deep ecologists will not differentiate in order to account for biological diversity and environmental complexity, relying on what amounts to a spiritual understanding of intrinsic worth in nature, and making human identification with nature an individual psychic act rather than a political practice (Bradford 1989:9, Pepper 1996:115). Whilst I do not share the hostility of some critics to the spiritual elements of some deep ecologism and ecofeminism, the need to account for relations of both similarity and difference between humans and other parts of "nature" remains. The human world is structured by systemic power relations that posit differential relations to the environment. The environment is biotically highly diverse and differentiated and acceptance of a hierarchy amongst the non-human environment is imperative if we are to posit a less distinct boundary between humans and the environment. This hierarchy may be biologically established in terms of species diversity and differential sentiency. It may involve the attribution of different kinds of value, different criteria for attributing it, and different treatment (Martell 1994:80), but this does not imply the acceptance of human domination. Respect for difference and diversity can result in non-dominatory differential relations with an incredibly diverse environment. "Differential value" involves respect for differences in type, form and interest, of the multivariate non-human natures. Deep ecologists need to account for differential social impact on the environment via sensitivity to the ways in which human-dominance over nature interacts and intersects with dominations of gender, race and class, amongst other complexities of difference. The deployment of theories of anthropocentrism in a way which juxtaposes human-centered and biocentric analytics and ethics cannot account for the complexity of domination in a world fractured by difference, just as a theory of patriarchy or of capitalism or, indeed, of a postmodem "condition," cannot by itself provide us with an inclusive theoretical framework.

Social difference and deep ecology

Within deep ecology there are conceptual possibilities for a consideration of intra-human difference. For example, Naess's concept of dwelling within "mixed communities" could enable an elaboration of the range of difference and a sense in which intra-human differences impact on specific formations of anthropocentrism. For Naess (1979:231), the mixed community referred to relations between humans and other animals. Others have used this concept in a manner reminiscent of Aldo Leopold's (1949:204) "land ethic" to include plant species and biotic systems of rivers, soils and mountains. According to Devall, "dwelling in mixed communities" implies adequate consideration of the needs, functions and potentialities of individual beings and their intrinsic worth, and the well-being of the community as a whole (1990:154). Current articulations however, ignore the historical, geographical and cultural context within which any particular "mixed community" has developed. This leads Naess to speak of a "racial prejudice against non-human life" (1994:106). Such slippage involves the implication of an absent referent (Adams 1990) – those communities of people who have incurred racist exclusion, discrimination and often violence. It behooves sophisticated thinkers such as Naess to conceptually distinguish the normative bias of anthropocentric society whilst being able to account for oppressive behaviors within and between human communities. Whilst there is no theorization of wealth and lifestyle, gender, sexuality, race, class, age and all the other crosscutting formations of difference, which are implicated in our relations to "nature," deep ecologists often distinguish between the city dweller and the more "authentic" inhabitants of rural areas. For Devall:

> The modern city is a necropolis – a vast city of the dead ... Communities of plants and animals are killed, bulldozed, burned and buried under human-created artifacts ... Only a few species, such as rats and some insects, seem to thrive in close company with humans but are not domesticated. (1990:188)

There is a deep-seated anti-humanism in a deep ecology that homogenizes our diverse interactions with the natural and "artifactual" environment. The simplistic denigration of the "city" above is deeply ironic. It is a result of deep ecology's inability to see that the notion of wilderness is a social construct at the same time as it might refer to relatively unmodified landscapes. All human communities are embedded in relations with nature, however distanced from relatively unmodified nature they might be.

Deep ecologism reflects elements of postmodern theory in the trenchant critiques of paradigms that characterized the development of modern societies. Yet deep ecologists often also adopt unreflexive modernist understandings of the "environment." They do not consider that "nature" or the environment is socially constructed, although they do argue the dominant Western conceptions of "nature" significantly shape how we treat the environment. Deep ecologists base much of their theorizing on an "essential" nature which can best be seen in the "wilderness," described as "relatively unstressed, untrammelled, and undisturbed habitat for wild species of plants and animals, that has been *protected and preserved*" (Devall 1990:163, my emphasis), which humans must refrain from "contaminating." Yet there is no wilderness unless, in anthropocentric society, decisions have been taken to preserve particular regions and localities. Such thinking belies the notion of human embedding in natural systemic relations, and pessimistically presents an amorphous humanity as a contaminating presence. Critics have referred to it as a form of primitivism which dates back to a golden age when, before agricultural and industrial revolutions, Western societies lived in some kind of harmony with their environments (Guha and Martinez Alier 1997:82–3). The narrative here of de-development reflects a deep-seated anti-modernism, yet the narrative of a fall from ecological grace is deeply and uncritically modernist. Sharon Doubiago (1989) saw entrenched sexism in the idealization of human–nature relations which stresses "wilderness" preservation, and of the experience of walking, running, cycling, even hunting in it, which is unlikely to be socially inclusive. Ariel Salleh (1984) has long contended that deep ecology ignores the material relation of women and men to production and reproduction, and the environmental consequences of such gendered relations, and more recently Salleh (1997) has argued that deep ecology makes social difference invisible.

We can understand a society as human centered, and see this as politically problematic without accepting the dichotomous deep ecological *strategy* of reversal – of the avocation of biocentric politics. I accept the conceptualization of human relationships with the environment as systematically and systemically exploitative, but deep ecologists can be simplistic and reductionist in arguing that this is a priori to other systems of domination. Deep ecologists do have an important point in stressing our embedded human condition, and the significance of our experiential relations with "nature" (see Anderson 2001), however, they need to consider the differential impact of human communities on the environment via sensitivity to the ways in which human dominance over nature interacts and intersects with dominations of gender, race and class. Without an analysis of intra-human domination deep ecological politics, although radical in tactics, tends to fall into a conservationism that does not question the basis of consumer capitalism which it so despises. The notion of our "dwelling" in mixed communities of difference may be a way of developing a more analytically complex and politically satisfactory deep ecology. I wish to deploy the notion of social relations with "the environment" as systemic and exploitative, but as existent in overlapping relations with other systems of social domination. First however, we must consider how other ecologisms have more successfully accounted for the interplay between human domination of nature, and our domination and exclusion of each other.

Social ecology: the problem of human domination

Social ecology sees environmental abuse and exploitation as a direct result of the domination of groups of human beings by other groups of humans – "intra-human domination." It does not conceptualize human dominance of the environment as an independent form of domination, but sees it as interrelated to oppressive systems of hierarchy amongst humans based on age, class, gender and race. Murray Bookchin is usually seen as the "founder" of social ecology (Roszack 1989), and his prolific range and extended reflection on humanities relations with "nature" make him the focus here, for as Andrew Light remarks, not uncritically, "Bookchin has risen to the task of filling in a complete body of theory largely by himself" (1998:9). Unfortunately, Bookchin has demonstrated an implacable hostility to critics from within left green and feminist circles. Yet his work constitutes *one variant* of social ecology, and socialist, liberation and feminist ecologisms are also best described under this umbrella. Bookchin is distinctive in the range of dominations he seeks to consider and is distinguished from ecofeminists and socialists by the strongly libertarian eco-anarchism of his thinking (Macauley 1998, Marshall 1993).

Systems of social domination and environmental exploitation

Bookchin has long argued that ecologism must involve an understanding of the causes and consequences of intra-human domination and oppression. The exploitation of humans by other humans is the key to explaining the human exploitation of the natural environment:

> the very concept of dominating nature stems from the domination of human by human, indeed, of women by men, of the young by their elders, of one ethnic group by another, of society by the state, of the individual by bureaucracy, as well as of one economic class by another or a colonized people by a colonial power. (Bookchin 1980:62)

One of Bookchin's most important contributions is the idea that humans are not equally responsible for environmental destruction. He has been intensely hostile to theories of anthropocentrism, considering the domination of nature as of secondary importance to intra-human domination (Bookchin 1990:44). In his most substantial work, *The Ecology of Freedom* (1991), he gives a complex account of how social hierarchies emerged with the oppression of women, proceeding to the exploitation and oppression of other groups of humans, socially stratified according to age, "race," class and sexuality. Bookchin argues these oppressions adopt different forms (being imposed by different social institutions and practices) and degrees of severity across different cultures and over time. In seeking to demonstrate how various social hierarchies emerge in interrelation, Bookchin draws upon anthropological literature in arguing that the earliest forms of human domination, led to the emergence of the domination of nature (1971:76). Yet as Alan Rudy (1998:275–7) has pointed out, much literature in pre-historical anthropology argues that the opposite is indicated. For example, in charting the development of patriarchal relations, Elizabeth Fisher (1979) has argued that the domestication of animals was key to the institutionalization of patrilinity. As with most stories of "origins," Bookchin is highly selective, and the data available is ambiguous and suggestive to a variety of interpretation.

Bookchin's analysis is both systemic and structural. For Bookchin, the understanding of natural biodiversity also applies to human society, characterized as it is, by extensive difference. Unlike natural eco-systems, social systems are marked by relations of hierarchy and domination (1991:29). Coercive social relations and practices of domination operate counter to the spontaneous and teleological evolution of life toward increasing complexity and consciousness. He sets up a false dichotomy here. All hierarchy in nature is non-dominatory, relations within ecosystems being mutual and reciprocal. In human "society" the converse is true. I think Bookchin is referring to difference here when he speaks of hierarchy, and as I argued in Chapter 1, it is not inevitable that social difference becomes implicated in systemic domination, nor that systems of domination are not also characterized by reciprocal practices. Difference amongst humans may be analogous to diversity amongst non-human species and within eco-systems, but it may also be implicated in relations of social domination. There needs to be an analytic distinction between difference per se and difference-in-domination.

Bookchin has, ironically, created an analytic hierarchy of oppressions, according analytic primacy to the state, attributed, rather strangely, to its ability to accentuate all social hierarchies. Whilst in earlier works Bookchin (1971) naively posited a move to a "post scarcity" society based on a superficial and overly optimistic reading of the dynamics of the capitalist system (Kovel 1998:47) in later works, Bookchin (1986, 1991) sees the operations of international capital as far more adaptable. His theorization of gender is ultimately, surprisingly minimal. Regina Cochrane (1998) argues that throughout Bookchin's vast historical coverage, women are depicted as passive onlookers to a male public sphere. Indeed, despite some critique of "male" anthropological gender bias, Bookchin himself is embedded in a dualistic understanding of gender (Rudy 1998). Whilst Bookchin can be described as a multiple systems thinker therefore, his theorization of certain systems of domination is inadequate.

What is missing is acknowledgment of the specific formations of social/nature – the ways in which human societies are embedded in natural systems and the varied formations and practices this implies. Bookchin's ecologism actually involves a relegation of what he has referred to as "merely" nature (Bookchin 1989). Antagonistic to any blurring of boundaries between humans and "the environment," Bookchin differentiates non-human nature as "first nature" and human nature as "second nature" by virtue of their "reason" and ability to interact reflexively with their environment (1991:xxi). Bookchin overemphasizes relations between differentially stratified humans at the expense of a more thorough analysis of society–environment relations, and has an unfortunate tendency to construct human difference from other animals as superiority. He homogenizes "the environment" rather accounting for the different ways in which and degrees to which, various aspects of "nature" are subject to domination, and the ways societies are embedded in and constitutive of, environments.

This said, Bookchin's contention that human domination and the domination of the environment are intrinsically linked is important. Both the domination of nature and of groups of human beings over other human beings in terms of class, gender, sexuality and race are produced by the construction of "Otherness" – the construction of groups of humans and the natural world as "Others" (Bookchin 1986:26). This fosters social structures based on dominance and submission (1986:55). I would argue that these hierarchical systems are separate systems of domination, whilst concurring that these systems of domination interlink and overlap.

Complexity and diversity

Bookchin has only recently come to endorse some of the concepts of the new complexity science, but I would suggest this has been latent in much of his work. Bookchin argues that webs of interdependent relations exist between natural phenomena and the eco-systems in which they are embedded (1971:58–60). This is similar to the arguments of the nineteenth-century geographer Peter Kropotkin. In *Mutual Aid* (1955), Kropotkin countered social Darwinism by arguing that the logic of natural evolution is not reducible to competition within and between species in order that the "fittest" survive, prosper and reproduce. Rather, mutual support, aid and defense are also significant factors in evolutionary adaptation within particular species. Similarly, for Bookchin, nature is unified despite its diversity, and species exist in relations of mutual interdependence and co-operation (1982:26). Bookchin considers hierarchy to be imposed upon human society, and like Kropotkin, thinks it is falsely attributed to the natural environment for political ends (1991:36). The concept of coevolution, which runs through *Mutual Aid* and *The Ecology of Freedom*, has been recently developed by complexity theorists in the earth sciences, as has Kropotin's representation of "life" in terms of multi-leveled and nesting systems, or "federations" of life forms. *The Ecology of Freedom* outlines an evolutionary model of human social development. Bookchin suggests social hierarchy emerged in the early Neolithic period with the establishment of rudimentary forms of government and the development of warrior groups to protect and extend territory. In his descriptions of evolutionary patterns and pathways, Bookchin considers that:

> The universe bears witness to an ever-striving developing – not merely "moving" substance, whose most dynamic and creative attribute is its ceaseless capacity for self-organization into increasingly complex forms. (1991:357)

Drawing on the work of microbiologist Lynn Margulis, Bookchin argues for symbiotic relations in "nature" between systems of land, sea and atmosphere, and forms of evolutionary cooperation/coadaptation. We participate in the evolutionary process, co-evolving with our environments and other species. However, whilst complexity science is not teleological, Bookchin's use of it is very much shaped by his Enlightenment narrative which tells of an evolution to a higher level of complexity and consciousness culminating in a state of "free nature" in which intra-human hierarchies are dissolved and the domination of the environment is no more. Bookchin bases his evolutionary hypothesis, from which much of his analysis of interrelated domination is derived, on a linear and in many ways predetermined model which is not bourn out by the scientific theories he draws upon in its development. Bookchin significantly underplays the elements of competition and predation in the evolutionary process, and sets up a dichotomous opposition between his own, and traditional Darwinian approaches, which does not have the nuance of Kropotkin's earlier synthesis. Bookchin suggests that increasing diversity and complexity in both nature and society is a source of ecological stability, and his understanding of that stability as dynamic and shifting echoes the approaches we examine in Chapter 3. Yet as Eckersley (1998) points out, Bookchin's particular view of evolution preserves a deep-seated division between humanity and "nature" and is highly anthropocentric, prescribing high levels of intervention in natural processes as highly desirable (Bookchin 1989:203). The problem is that the possible results of human intervention are, given the complexity of natural systems, difficult to so divine, and an approach and practice which emphasizes mutuality and diversity rather than an objective "logic of evolution" is more appropriate.

Part of the problem is the lack of reflexivity in Bookchin's account, his somewhat dated anthropological and system science sources, and his loose analysis of social hierarchy. Glen Albrecht (1998) has argued that some of the key arguments advanced and substantiated by complexity theorists can counter many of the criticisms made of Bookchin. Yet it was only relatively recently that Bookchin (1990) explicitly referred to the work of complexity theorists such as Stuart Kaufman and Ilya Prigogine. In Chapter 3 I argue that complexity theory can be usefully synthesized with elements of ecologism, including many of Bookchin's ideas. Yet the inadequacies and contradictions in Bookchin's theorization of difference mean also that his deployment of complexity approaches does not facilitate the complexity of analysis that it might.

Social ecology and different natures

In contrast to deep ecologists, Bookchin adopts a hierarchical approach to environmental ethics. Humans are "second nature" and as thinking and reflexive beings, they can act as the "voice" of first nature – the ecosystem (1990:182). The environment only has "rights" if, when and how, human beings decide to confer them, whereas humans have intrinsic value by virtue of their powers of reason and creativity. This uncritical stance on Western modernity is problematic, for as Val Plumwood (1993:15) argues, the elevation of human reason has been a key justification for human domination. Yet Bookchin argues for an ethical position that understands forms of hierarchy in nature that do not imply domination, proposing an "ethics of complementarities" (1991:xxxvii) based on degrees of differing "sentiency" (cognitive awareness, ability to experience pain and pleasure), to account for variety amongst beings, which are highly differentiated in terms of biotic development. This avoids the problem of the homogenizing ethical tendencies of some deep ecologists, but takes us too far in a humanist direction. Hierarchical relations of domination apply only to human society with organized, systemic practices and institutions, and Bookchin would see such an application of dominatory hierarchy to the environment an "anthropomorphic projection" (1991:xxiii). Here lies the crux of the problem with his position, a fundamental divide between humans and other species.

Bookchin is mistaken in seeing humans and all animals as incomparable categories. I want to combine Donna Haraway's (1991) burring of the harsh boundaries between humans and all other animals, with Bookchin's understanding of a hierarchy amongst animals, wherein human relations with other life forms should be based on an appreciation of difference in terms of complexity and sentiency. I would depart from both positions however, in arguing that the social construction of "nature" imposes relations of dominance upon all elements of the environment, which although vastly differentiated are homogenized as Other, in a society organized around the principle of human domination.

Ultimately, Bookchin's social ecology overemphasizes the impact of intra-human domination as causal in environmental degradation. The key point of deep ecologism is that the society–environment relation has been marginalized in social and political theory, and that such a relation is a form of domination with devastating consequences not only for humans, but most immediately, for plants, the diversity of non-human animal life and natural systems. In the scope of his coverage of intra-human oppressions, Bookchin does not analyze each in sufficient detail, nor elucidate exactly how any one form of domination results in certain forms of environment–society relations. "Bookchinite" social ecologists have often been hostile to other positions (Biehl 1988, 1991). This is a pity, because some ecofeminist and eco-socialist analyses could be usefully deployed in developing Bookchin's ideas on the interconnecting forms of social hierarchy in relation to the environment. Other social ecologists, such as Theodore Roszak (1992), have developed approaches that combine the rationalism of social ecology with an understanding of humanity as part of, and continuous with, the environment. Marxist and socialist ecologists with respect to the impact of capitalism, and ecofeminists with respect to gender relations, have more successfully accounted for specific forms of social domination and the domination of the environment. A narrative of hysteria in which any alternative radicalisms are caricatured and pilloried, mars Bookchin's work (Clark 1997:9). What makes this so unfortunate is that in many ways, Bookchin sets an agenda for a social ecology that can involve (multi) systemic analysis of various forms of social domination, a systemic approach to the human domination of the environment, and a complexity understanding of systemic difference-in-domination. It is perhaps the ambition of this project, which is both a strength and a weakness here. I want to use an overarching concept, and domination plays a similar role for me as hierarchy does for Bookchin. However, whilst we need concepts that draw together different systemic analyses, we also need to allow conceptual relative autonomy in order to elaborate specific formations, which the dominations of difference assume.

Eco-socialism: capitalism and the commodification of nature

Historically, Western socialism has had little to say about the environment, and when it has, it has shown little sympathy toward environmentalist perspectives, bar the latter work of some members of the Frankfurt School (such as Marcuse 1972). Since the early 1980s a number of socialists and Marxists have explored the relationship between "red" and "green" thought and concerns and like eco-feminists and social ecologists, have been interested in the relationship between social inequality and difference amongst humans, and the domination of "nature." Similarly, eco-socialism is eclectic, housing a range of socialisms and ecologisms, and distinctions between some kinds of liberation ecology and eco-socialism are blurred at best.

Some eco-socialists have argued it is the specific use which capitalism makes of industry not industrial production per se which constitutes the problem. Such use is directed, according to Joe Weston (1986:4), toward the creation of profit rather than the satisfaction of social and economic "need," and the consequence, poverty, is the main cause of environmental problems. It is not clear however, what is specifically ecological about such theorizing. Weston does not critically evaluate concepts of wealth and of need, and leaves unanswered the questions raised by deep ecologists about the unsustainable nature of Western affluent consumption driven societies. Others focus on "communitarian" socialism and contend both that this has much in common with, and is a necessary grounding for, ecologism in providing an understanding of social and political life that cannot be analyzed in the same ways as human–environment relations (Ryle 1988:7,20). Since the early 1980s, some eco-socialists have utilized some of the concepts of classical Marxism in order to understand environment–society relations. Elements of "classical Marxism" are potentially useful in developing "green" social theory, eco-socialists argue, and in doing so, they have deployed systemic analyses of capitalism.

The legacy of Marxism: greening an ontology of capitalism

Peter Dickens (1992) has argued that Marx has an important contribution to make to the understanding of society–environment relations. Marx has a dialectical conception of relations between society and the natural environment, wherein humans depend on nature for their existence, and are shaped by and in turn alter, their surroundings. In addition, we are dependent on the natural world for the realization of our intellectual and aesthetic powers. Ted Benton (1996), for example, has emphasized the concept of "species being" in the early writings of Marx:

> Species-life, both for man and for animals, consists physically in the fact that man, like animals lives from inorganic nature. ... Man lives from nature, i.e. nature is his body, and he must maintain a continuing dialogue with it if he is not to die. (Marx 1955:327)

Benton's (1989, 1993) eco-socialism is developed around these key themes in Marx of natural limits and human nature (species being). He has made a powerful case (Benton 1993) for a reassessment of human–animal relations taking into account the diversity and specificity of non-human animals specific "species-being" in relation to our own. For Marx himself, labor was key to human species being, and Dickens (1996) suggests it is the social organization of labor power that is the key influence on environment–society relations. Of necessity, we work on nature to produce the things we need. Certain ways of organizing labor in capitalist societies around the production of goods for the market on the basis of increased production and consumption to satisfy the profit motive means that the natural environment is exploited.

David Pepper also emphasizes that humans constantly interact with nature and thereby change it (1993:111) and this is both material (physical change to the environment, forms of human labor power and technological development) and ideological (it influences how we think about nature). We are in a dynamic dialectic relationship with the environment, he charges, changing nature as it is changing us. As economic globalization means industrial manufacture moves to the poorer countries of the world where labor is cheaper and pollution controls fewer, the links between industrial production, environmental damage and exploitation of workers are perhaps stronger now than when industrial manufacturing was at its peak in affluent locations. The notion of the dialectic enables us to conceptualize a two-way flow of change and causality in which the environment has its own properties and causal powers. For some, the contemporary development of capitalist relations means that "nature" becomes increasingly internal to the dynamics of capital accumulation (Castree 2001:191). Noel Castree argues that agriculture has posed problems for capitalist development because it has been incredibly difficult to physically alter "nature" to enhance profit (seeds, for example, reproduce themselves). In contemporary times capitalist corporations have harnessed biotechnology in order to overcome "natural" barriers to capital accumulation (by manufacturing hybrid seeds, or those which grow into sterile plants). Following a similar line of argument, David Harvey suggests that industrial capitalism has meant that we now have what he calls "constructed ecosystems" (1996:187), and as such, the human production of natures is so established, it cannot simply be reversed. There remains a tendency to argue that environmental exploitation is not systemic within socialism as a form of social and economic organization however. James O'Connor (1989) contends that whereas capitalism as a system of economic production premised on the profit motive is inevitably polluting and depletes natural resources, socialism, based on the principles of reciprocity and redistribution, does not have a built-in propensity to pollute. Pepper (1993) argues that an eco-socialist mode of production will produce and distribute goods and resources according to need not the demands of capitalist consumerism (1993:146). O'Connor and Pepper assume that given the limits on "need," a socialist system of production in contemporary society would place less pressure on natural resources and be better placed to operate within natural limits.

Kate Soper (1995) is right to point out that anthropocentrism remains endemic in Marxism, which tends to view nature as a series of resources for human ends. Even "green" Marxism is problematically complicit in the Enlightenment narrative of human dominion. The emphasis on "natural capital" – the drawing of nature into the productive process, is problematic for the production of nature predates capitalism (Thomas 1983). Chapter 6 contends that Marxist concepts are crucial to understanding the ways in which nature is produced/reproduced, but these processes are also embedded in and constituted through gendered and natured domination. Capitalism as a system of economic production and social relations alienates us from the natural environment and from ourselves. It encourages us to see the natural world as a series of commodities for use rather than things we work with in reciprocal relationships. However, to borrow Castrees's (2001:191) words, Marxism is a "necessary but not sufficient" approach to understanding the complexities of human engagement with and exploitation of the environment.

Social inequality, the degraded environment and class politics

Eco-socialists have been concerned with a range of forms of social difference in terms of the analytics and politics of social exclusion. In particular, they concur that social inequality based on class affects the quality of the environment in which we live, so for example, in affluent regions of the globe, it is the poorer sections of the working class that suffer unhealthy jobs and polluted living environments. Pepper (1993:63) argues that much socialist political action in nineteenth- and twentieth-century Western societies involves what he calls "environmental protest" in the forms of struggles against conditions of life in factories and the ecological dislocation of mass industrialization and prerequisite urbanization and rural restructuring. Pepper (1993:3) then insists that "human rights" provided for by socialism are prerequisite for developing a more benign relationship toward the environment. Whilst some eco-socialists have theorized in terms of a diverse selection of socially marginalized groups in a post industrial context (Gorz 1982, see Scott 1990), others see any marginalization of the political significance of class (Pepper 1993), or move away from a mono-systemic analysis (Harvey 1990:46) as some kind of postmodernist drift toward "lifestyle politics." Marxists are right to emphasize the iniquitous distribution of environmental bads in poor communities (locally and globally), and the environmental implications of the global division of labor and the exploitation of workers in underdeveloped countries by transnational corporations. Yet an understanding of the systematic dominations, exploitations and oppressions of capitalism must be nuanced by taking difference more seriously.

Social ecologists such as Roszack and Bookchin, and eco-socialists like Benton, include gender, but analyze gendered ideas and forms of environment–society relations in insufficient depth. Pepper manages to dismiss eco-feminism as a "mishmash of mysticism, morality ... and (the) division of the world into 'all that is good is female, all that is bad is male'" (Moore 1990, endorsed by Pepper 1993:148). Synthesizing elements of socialism, ecology and feminism, Salleh (1997) argues that we must recognize the socially constructed division between humanity and "nature" at the same time as we question divisions and exploitations based on gender, "race" and class. Salleh gives a broad account of the interconnecting webs (Plumwood 1994) of social exclusions, exploitation and oppression that shape the differing relations of human societies to their non-human environments. Central to materialist understandings of the world has been the notion that nature is both produced and exploited. In contemporary times, this perhaps has greater resonance in the context of an ever more effective technological domination of nature closely linked to globalizing processes, powers and institutions of capitalism.

Different worlds, different problems: environmental justice and liberation ecologies

Deep ecologism has been emergent in particular places: the United States, Canada, Australia and some parts of Scandinavia, where environmental organizations have tended to focus on wilderness issues, and there are significant tracts of land relatively unaffected by human patterns of land use. This has led critics to describe deep ecologism as an exclusively Western phenomenon (Guha and Martinez-Alier 1997:xiii). In parts of the Southern hemisphere the links between environmental degradation and intense forms of poverty are very strong, and debates about the non-human environment are bound up with those around "development." It is within academic debate and political activism around "development" that liberation ecologism has emerged, and its insights have been applied more broadly in the political language of "environmental justice." Liberation ecologism suggests that struggles for intra-human justice are closely bound with those to prevent environmental exploitation. In theoretical terms, it brings the conceptual apparatus of postcolonialism to bear on debates around social difference and human–environment relations. In some ways I am uncertain that "liberation" or anti/postcolonial ecologies are best considered separately from feminist and socialist accounts. Certainly in origin, liberation ecology has emerged fairly directly from Marxist and neo-Marxist political economies of development (Peet and Watts 1996:x), and some theorizations are clearly Marxist or socialist feminist in terms of the conceptual apparatus deployed (Blaikie and Brookfield 1987). I attend to "liberation ecology" as a distinct strand of ecologism because within much of the literature on green social and political theory, anti-postcolonial accounts of society–environment relations are all but absent. Yet liberation ecology is not the "newest" shade of green. Ramachandra Guha and Juan Martinez Alier (1997) for example, make a powerful case that Mohandas K. Gandhi is a founding figure of a "social ecology of the poor," with his critique of the interrelated social and environmental impacts of industrialism, Western capitalism and the imperialism of Western style development.

In addition, liberation ecology might best be seen in the plural. Like ecological feminisms, there are a number of epistemological and ontological positions adopted by liberation ecologists. From the early 1990s, post-structural concerns with cultural difference and the deconstruction of knowledge-as-power have moved some liberation ecologists toward discursive and postmodernist understandings of development and society–environment relations (Slater 1992, Escobar 1996). Richard Peet and Michael Watts (1996:13) make engagement with poststructuralism part of their definition of liberation ecologism but I consider this stricture to be undermined by the range of positions within such theorizing.

The colonialization of nature

Postcolonialism has been used to identify a historical period and the patterning of continuous relationships between the former colonizers and colonized. More often, it has been an interrogation of such interrelations from the inauguration of the colonial encounter (Said 1978). European cultures of nature, which emphasized domination and domestication were practices applied to colonized peoples who were "naturalised." Richard Peet (1985) has shown that during the early twentieth century for example, native peoples in colonial territories were understood as "primitive," "backward," "uncivilized" and a key element of such racialized discourse was a notion that such people were both closer to nature and naturalized by being "closer" to an animal state from which white western men had transcended. Colonial institutions and practices sought to demonstrate their domestication via "stagings" of the power to dominate nature (such as dams, canals, plantations), which as Ramachandra Guha (1989b) notes could be disrupted by political protests and unforeseen physical events. "Nature" was a means of registering difference within colonial discourse, and the notion of "tropicality" represented colonized nature as abundant or pestilent (Arnold 1996:140–60). Such pathologisms were applied to colonized peoples, and, as Derek Gregory (2001:100) contends, cross cut by racialized and sexualized discourse of the "natural."

Whilst the public discourse may be moderated, the material practices of postcolonialism constitute more of a direct continuum. According to Guha (1989a) the environmental policies of "developing" states in the postcolonial order were derived from European and American conceptions and practices such as "rangeland management" and the establishment of national parks. This has led, for example in the case of the tiger "reserves" in West Bengal, to the clearance of peoples from ancestral lands, legitimated by a discourse of conservation imperialism (Guha 1997:xvii). Much of the environmental debate in the South hinges on how the process of modernization and development is managed and controlled, with Northern control of the development seen to be the problem rather than industrial development per se (Chatterjee and Finger 1994:77). Environmental difficulties are very much embedded in the social relations of colonial capital, but those specific problems differ across the global South. For Peet and Michael Watts (1996:14) different situations give rise to different regional discursive formations of environment–society relations. In this analysis, the complexity of difference is best captured not by overarching theorizations of, for example, anthropocentrism as advanced by deep ecologists, but by attempting to grasp the specific forms and articulations of environment–society relations which are hybridities of social nature, and at the same time, vastly different according to geographic specifics. I would suggest that we can have both an analysis of specific forms bounded by time and place, whilst also considering these formations to have been produced by the particular effects of overlapping or interpenetrating systems of social domination.

Environmentalism(s) of the poor

Guha (1997:23) has argued that the "environmentalism of the poor" differs from the environmentalism of those in wealthy parts of the globe in that it is more concerned with social justice, and tends to arise from distribution conflicts over resources linked to livelihood. Poor communities may respond to environmental damage which directly impacts on them, but do not share certain Western preoccupations such as "mere" protection of species and habitats. In addition, third world environmentalists argue that the poor regions of the globe have to deal with more severe environmental hazards than the relatively wealthy North, and the notion of environmental justice also involves the defense of human rights. The well-known cases of Chipko in India and the rubber tappers in Brazil to preserve forest eco-systems, or the Ogoni people in Nigeria against pollution by Shell Oil, are all environmental struggles embedded in the politics of justice for dispossessed, displaced or disrupted human communities. Consequently, Guha argues that fragmentation and uncertainty is characteristic of "third world" environmentalism, and that this implies a lack of coherency and abstraction in the environmentalist thinking of the poor. Unlike first world ecologisms, those of the poor are more likely to be what Haraway (1991) would call situated knowledge's emergent from localized condition and specificities. However, the "environmentalism" of the poor is not so simply geographically clustered. Robert Bullard (1990) argued disproportionate environmental hazard was experienced by poor black Americans, and advocated environmental justice as practice and theory. Significant international Nongovernmental organizations (NGOs) emergent from a "first world" context, such as Friends of the Earth are now actively pushing a social ecology agenda, and the influence of postcolonial theory has been incredibly important in encouraging ecosocialisms and feminisms to foreground their theorization of ethnicity and locality.

Liberation ecology strongly asserts the embedding of communities within environments, and is anthropocentric in that it takes as read, in a similar way to eco-socialists, the "rights" of communities to utilize natural resource, but it is careful to differentiate different kinds of appropriation of "nature." Guha characterizes environmentalism in poor countries as conflicts between those communities dependent on the natural resources of their locality – "ecosystem people" and "omnivores" – individuals and groups with the social power to transform and use natural resources on a scale beyond the locality who have potential to exploit such resources. What is useful about such a conception is that it brings together some of the insights of deep, social and socialist ecologies, drawing on bioregionalism, the social embeddedness of human "society," and class analysis.

Despite their critiques of "Western" ecologisms, Guha and Juan Martinez-Alier adopt a systemic perspective, which places the diversity of human communities in the context of both diverse "physical–chemical–biological" systems and within the global capitalist economic system (1997:22–7). They have suggested that what is most critical to ensuring environmental justice would be the "de-development" (1997:xx) of the first world, for wealth is a greater "threat" to the environment than poverty (1997:59). The West must abandon the eco-capitalist notion of "sustainable development" and consider that the over development of the "West" has been part of a process of deeply exploitative colonial and postcolonial relations. This challenge to conservation imperialism is also a challenge to international capital in the form of transnational corporate practice and accumulation. However, the characterization of the environmentalism of the poor as localized, social, community based as against the environmentalism of the rich is not clearly sustained. Whilst international environmental NGOs undoubtedly engage in "environmental imperialism," and deep ecology in particular does not acknowledge the situatedness of its preoccupations, the environmentalism of the rich is highly fragmented and diversified in both theory and practice. There are important differences between ecological perspectives of rich and poor regions and communities, but to argue that there are "two environmentalisms" and that one is that of affluence and enhanced quality of life against another of survival and livelihood is both to oversimplify and dichotomize such differences, underplaying the complexity of relations of race, gender and class and other formations of differences which enable divergent emphases of various kinds of social ecology.

Postmodern liberation?

In the early 1990s emerging theories of liberation ecologism tended to draw upon neo-Marxist material on third world "underdevelopment" (Peet 1991) considering the depletion of natural resources to be endemic in the capitalist world system. Yet those identified with liberation ecologism have more recently suggested that Marxist influenced theorizations are often reductionist in positing poverty as the (mono)cause of environmental problems in the third world and in being preoccupied with the labor process, evade the specific dynamics of power in ecologically embedded communities (Peet with Watts 1996:7–20). Some have been drawn to poststructuralism in arguing that different regions of the globe have different ways of thinking about and acting upon the issues of "development," modernization and environment that cannot be considered within a model of global capitalism.

Arturo Escobar has perhaps moved ecologism furthest in a poststructural direction, drawing on discourse analysis in order to understand the social construction of nature. Whilst for Escobar, nature is both known and "produced" through discourses and social practices, he suggests this is "something entirely different from saying 'There is no real nature out there' " (1996:46). Nature is a multiplicitous discourse with real existence and effect, which can be constitutive of a materiality of nature. In deconstructing the notion of "sustainable development," Escobar (1996:49) contends that this environmental discourse actively shapes people's perception of nature and is political – promoted by a first world led coalition of governments, business organizations and scientific authorities. Escobar utilizes both a systemic analysis of capitalism alongside that of discourse and narrative. He acknowledges the deleterious impact of modern capitalist relations in the "third world" whilst unpacking the ways in which capital deploys "expert" scientific discourses. He links "biodiversity prospecting" of commerical biotechnology to the establishment of localities of biodiversity over which indigenous communities are required to be in "sustainable management" (1996:57). So Escobar broadens the analysis to include the contestation of environmental discourse in a postcolonial context, as well as commenting on the resource appropriations of capital. An important element of the colonial project was the cultural colonialism of knowledge particularly the spread of Western notions of rationality. Discourses of development, sustainable or otherwise, are understood by Escobar as hegemonic discourses reflecting postcolonial relations which subject peoples of the third world to Western power-knowledge, and increasingly ensure conformity of peoples and communities to the economic, political and cultural practices of the first world.

Whilst Escobar invokes the language of postmodern analytics in speaking of the "postmodern form of ecological capital" his theory is indebted to eco-socialist analyses of the increasing exploitation of "natural capital." This exploitation is coexistent with the practice of sustainable development, and Escobar sees such duality of practice as postmodern. What he describes here however, is the coexistence of old capitalism within the recent processes of the greenwashing of capital accumulation. Escobar (1994) moves further toward a fully postmodern approach in endorsing Haraway's characterization of the hybrid objects and organisms of contemporary culture as "cyborg" and concurs with her evaluation of contemporary society as postmodern and post industrial, wherein the "dissolution" of the society/nature split is a real possibility (Escobar 1996:61). There is of course, a distinctly modernist teleology posed in this idealistic transition from modern industrialism to postmodern postindustrial capitalism. To accept the hybridity of nature does not imply some kind of desirable political change, and to suggest that it does, as does Escobar in following Haraway, is utopian. The notion of coexistence between a first world characterized by postmodern ecological capitalism and a third world dominated by an essentially modern capitalist regime of "sustainable" development is not substantiated. Agriculture in first world countries for example, is highly industrialized, with significantly deleterious ecological effects.

Like ecofeminism, liberation ecologism is rather more ontologically and epistemologically eclectic than social, socialist or deep ecologisms. Like ecofeminism and deep ecology, liberation ecologisms problematize Western modernity, particularly the deployment, cultural and political, of rationality. Marxists, post-Marxists, Foucaultians and more postmodern poststructuralists can all be seen to be engaged in the theoretical production of liberation ecologisms. This said, I am in agreement with Piers Blaikie (2001:147) who argues that all the differing positions within the gamut of liberation ecology cannot "abandon nature to an unresolved relativism, where one view is uncritically accepted alongside many other, often contradictory ones." Liberation ecologists can only move as far as a soft constructionism, and even the work of Escobar is, in important ways, compatible with the critical realism I advocate in Chapter 3.

Those liberation ecologists who move furthest in the direction of postmodern theory often intimate some realism – by invoking chaos theory in the natural sciences as "evidence" of the uncertainties of "nature" and its integration with the social. And yet, no modernist social ecologist/ecosocialist/ecofeminist would disagree. However, postmodern theorists have appropriated "chaos" theory as a new scientific (phenomenological) epistemology and described such theory as legitimating their picture of a fractured, uncertain, world (see Escobar on Maturana and Varela 1996:64). The ecology of chaos (of disequilibria, instability and chaotic fluctuation) is set up as evidence of the outdated systems theories, which are seen to emphasize stability, harmony and resilience in the environment. Such an understanding assumes those systems, both natural and social are conceptualized as equilibraic and predictable and that systems analysis simplifies the variety of this planet. In contrast however, developments in the science of chaos have led to the notion that within the chaos of nature, complex systems are emergent. I draw on developments in complexity theory in Chapter 3, and integrate these with the strengths of various ecological approaches I have reviewed here.

Deep ecology has provided social theory with a vital challenge: the possibilities of theorizing beyond and across the culture/nature divide. Theories of anthropocentrism have incredible political and analytic importance in making a case that our social systems are structured in ways that are predicated on the exclusive centering of human needs. I concur that there is a social system of human domination, but consider that like all systemic social domination, this takes historically and geographically specific formations. Bookchin's social ecology has been immensely important in presenting environmental domination as intrinsically linked to multiplicitious intra-human formations of domination, oppression and exploitation, yet the precise form of these systems, the detail of their interrelationships and the specific way they shape regional, global and local formations of socialnature, is not developed. What I see as other forms of social ecology: feminist, socialist and liberation ecologisms have fleshed out the ways in which capitalism, patriarchy and postcolonialism have had specific effects upon and given rise to, particular formations of socialnature. No social ecology however has argued for bringing together analyses of these varied relations of human domination within both a multiple systemic perspective, and one which integrates and analysis of human domination as a social system, with those based on the differences of human domination.

What is required is a social ecology of complexity that examines intermeshing, multiple systems of relations of social domination and considers them to be implicated in and constitutive of, the human domination of the "natural environment." In order to develop this, Chapter 3 argues the case for a critically realist epistemology and examines the relevance of complexity theory for understanding both natural and social systems and importantly the overlaps and synergies between multi-leveled systems of socialnature. Chapter 4 focuses on feminist theory and argues that theories of social difference and domination have often been reticent in theorizing beyond the social. Some have engaged with systems approaches in understanding gender as webs of relationships of domination and problematic difference however, and I argue that certain feminist approaches allow for cross cutting social difference by multiple systems analysis. Despite this, conceptualizations of the human domination of nature are rarely integrative of feminist theorizing, despite the greening of certain theorists work. Chapter 5 will make the case for ecofeminism as potentially rich for the development of such an approach, but considers that the lack of systems theory has been a hindrance in capturing the complexity of interrelations of domination with which ecofeminists have been concerned.

3
Complex Systems: "Nature," "Society" and "Human" Domination

> The species divide is not *solely* a behavioural or biologically-determined distinction, but a cultural and historically changing attribution ... humanity has persistently been seen *not* as a species of animality, but rather as a condition operating on a fundamentally different (and higher) level of existence to that of "mere" animals ... the norm of the human became identified with the achievements of "civilized" Western humanity measured (especially under modernity, but also for the ancients) in terms of acquiring technical control over nature.
>
> Kay Anderson (2001:80, my emphasis)

"Nature," "society," and "human" – all are highly contestable terms. Increasingly it is becoming accepted that nature is social, and as such, variably constructed across time, space and place (Soper 1995, Cronon 1995, Mcnaghten and Urry 1998). Some sociologists are questioning prevailing conceptions of "society" (Urry 2000) and more radically, suggesting that sociality is not exclusively human (Benton 1993). Yet as Kay Anderson suggests above, the "human" is also a social construct linked to formations of power, and all ecologisms would agree, albeit in different ways.

Biologists such as Lynn Margulis and Dorion Sagan (1986:214) have argued that there is "no physiological basis for the classification of human beings into their own family." Such taxonomy is reflective of anthropocentric bias – we are, they say, great apes. There can be no doubt as to the radicalism of this claim. Controversial utilitarian philosopher Peter Singer is well known for his advocacy of "rights" for animals, and most recently for his involvement with the "Great Ape Project" – the attempt to include chimpanzees, gorillas and orang-utans within the auspices of the UN Declaration of Human Rights. If the firm boundary between "humans" and other great apes were ruptured in this way, the implications would be profound. Yet again, we would have to question our means of distinction and formations of social power in the world. The power of wealthy, white, western men in modernity has been constituted by and through constructions of class, race and gender. These categories may be separate or mutually constitutive – they are overlapping referents, implicated in networks of difference. Also, however, these social categories of difference and domination have been natured. To say that class, race, gender or nature are "socially constructed" does not mean that they do not have real effects, or that they do not have a physical referent. Prevailing ideas about "nature" have implications for the treatment of certain categories of humans who are "natured" and have certainly impacted non-human species of animals, plants and their contexts. These impacts might be seen as effects of systemic human domination, on the construction of nature(s) with the referent of species. Humanity might best be exposed as a construct, ironically, by referring to the similarities of physiology that render this taxonomy political.

This chapter sets out the framework by which I draw upon the different ecologisms outlined in Chapter 2 and argue for a social system of human domination. I suggest that this exists in a milieu of multiple systems of social domination, one of which is gender domination, the subject of Chapter 4. I advocate a "complex systems" perspective for theorizing a whole range of interlocking relationships: those between "environment" and "society," the webs of relations between plant and animal species, between species and the contexts in which they are embedded, between different groups of humans and various groupings of species and context. In arguing for such a perspective, I need to engage with some well-worn epistemological arguments. The divisions between constructionism and realism, whilst sometimes over-emphasized, are not arbitrary, passé or surmountable. I consider some form of realist epistemology necessary for the development of systemic theorizations of environment–society relations, and I argue for critical realism here, and link it to certain themes and conceptualizations from complexity theory.

Complexity theory emerged from chaos theory in the natural sciences. The latter has been "appropriated" by postmodernist accounts of social life, through a reversal of the scientific understanding of chaos as the occurrence of complex information in which order is emergent (Hayles 1991:176). Common to both postmodern social theory and "scientific chaos" is the rejection of mastery (Hayles 1990:292), but the claim that realist scientists are absolutist and unreflexive is not substantiated. As James Lovelock (2000:11) puts it so nicely: "The practice of science is that of testing guesses; forever iterating around and towards the unattainable absolute of truth." Thus scientists usually acknowledge the provisional qualities of their knowledge "without the benefit of the deconstructive zeal of social scientists!" (Blaikie 2001:143). Some would argue that the provisional and qualified nature of complexity theory renders it compatible with postmodern approaches (Cilliers 1998), others are of the view that the breadth of complexity theory enables us to transcend many of the polarities and tensions in social theory (Walby 2003a). I am sympathetic to David Byrne (1998) however, who considers that the ontological and epistemological implications of complexity theory make it part of the realist program of understanding and enquiry. For example, a key element of critical realism is the notion that causality, both in society and "nature" is complex and contingent, and this is entirely commensurate with a complexity understanding of systems exhibiting different emergent properties and powers. Thus specific properties and powers differ at different levels of analysis.

Complexity thinking, originating in the "natural sciences" has produced a range of new ways of understanding the notion of "system." It suggests that we might capture the complicated yet organized quality of social and "natural" systems by understanding any system, as existent in a milieu or "environment" of an enormous range of other systems. Whilst systems are distinct, they are interlinked (with two-way flows of interrelationships), and importantly operate at different levels – thus systems may be "nested," exist within other systems. Systems thinking is significant for critical social and political theory because it is concerned with networks of relationships, and the embedding of such networks in larger networks again. As Fritjof Capra puts it "For the systems thinker, relationships are primary" (1996:37). So too, as we saw in Chapter 2, are ecologisms crucially concerned with relationships. Deep ecologists ponder the patterns of species relations, relations between peoples and their dwelling places and more grandly, of humanity to the "environment." Social ecology maps the complexities of intra-human domination to changing relations between environment and society. Liberation ecologies presume that human–environment relations are socially embedded, and like eco-socialists, would consider poverty and multifaceted social exclusion to be systemically constitutive of different formations of environment–society relations. Whilst the diversity of feminisms have been concerned with a plethora of gendered relations of difference, ecofeminism has opened itself up to a consideration of non-human difference and domination, and the relations between the latter and social formations of gendered, sexualized, racialized and class-based inclusion/exclusion. Through a complexity frame, we can engage with the difficulties of complex difference whilst developing a systemic approach to social domination.

The introduction outlined a particular definition of "nature" and "environment." The environment is conceptualized as the non-human animate world and its contexts – the whole range of multifarious animal and plant species, soils, seas, lakes, skies. Whilst there are incredible differences between and amongst these phenomena, they are grouped merely by biological referent: being both non-human, "live" (manifesting properties of metabolism, growth, reproduction, response to stimuli) and/or being part of an ecosystem. In societies structured around relations of human domination, the complex and highly diversified non-human animate lifeworld is homogenized as "Other" to the human, and often referred to as "nature." The construction of this Other is political. "Nature" is a socially constructed category based on power relations, and is manifest as a dichotomy between human beings and the environment. My chosen concept for the systemic domination of nature is *"anthroparchy."* I use it to describe a particular modern formation of social relationships in which non-human nature is cast as a series of resources for human ends, and in which human interests organize the systemic ordering of social control over the environment. Within and despite such relations of social domination however, the causal powers of the environment are manifest in the emergent properties of both relatively unmodified and human-modified eco-systems, which operate within/across/alongside anthroparchal networks of relations.

Theorizing "nature"

Social theories about nature have been shaped by different disciplinary preoccupations and histories. For example, recent years have seen a profusion of works within critical/cultural human geography, which have attempted to de-naturalize nature, that is, to argue that nature is socially mediated and constituted, and to problematize the boundaries between the human and non-human worlds (Whatmore 1999:23). Critical geographers have tended to argue both that nature is intrinsically social, and there is a plurality of knowledges about it, whilst also being in agreement that "the idea that nature is nothing more than a social construction" is "outlandish" (Castree 2001:16). Many have learnt to live with an uncertain knowledge of nature/environment, whilst also considering that some human practices are "better" or "worse" for the environment. This "soft" social constructionism is an element of the critical realist approach, but we also require an understanding of non-human nature as having emergent properties and powers that may be temporarily or permanently beyond our knowledge and experience.

In environmental sociology, disputes between constructionism and realism have been protracted, heated and remain unresolved. Sociology has a legacy of studying human society as separate from the "environment," which it tended to define as apart from the social world (Macnaghten and Urry 1995:203–4). Ted Benton and Michael Redclift (1994:3) see this as a result of the debates surrounding the development of the discipline in the early twentieth century. Sociologists insisted on the distinctiveness of the "social" from the "natural" as a means of countering the influence of biology in the explanation of social phenomena. Benton (1998) has described this legacy as "nature-phobia." Riley Dunlap and William Catton have long argued that the dominant sociological paradigm is "human exceptionalist" (1980:34). This conceptualizes society as exclusively human, and needs to be replaced by a "new ecological paradigm," which considers the impact of environmental factors on social behavior, and the impact of social processes on the environment. By 1993, Dunlap and Catton concluded that environmental sociologists persisted in operationalizing the traditional human-exemptionalist paradigm by adhering to a strong form of social constructionism. The exchanges between constructionists and realists have tended to involve caricature and a lack of compromise. In a relatively sober discussion, Kate Burningham and Geoff Cooper (1999) suggest that the differences between each "camp" are not as marked as the protagonists assume, but nevertheless, they "emphasize the appeal and utility of social constructionism" (1999:297), and this has been the over-riding sociological tendency.

Social nature

There are marked differences in what it means to understand nature as a social construction, and what implications arise from such understandings (Braun and Wainwright 2001:41). Those attracted to postmodern conceptions have developed "strong" constructionist understandings of "nature," reflecting the postmodern condition of fragmentation and refraining from universalizing "grand" theoretical schema, which are seen as incapable of catching the complexities of social life (Lyotard 1984, Lash 1990). Postmodern approaches have tended to focus on subjectivity, meaning, and cultural process, refraining from the deployment of systems analysis, and dismissively deconstructing knowledge produced by the environmental sciences (Wynne 1994, 1996). John Hannigan (1995:3) argues the "scientific establishment" constructs environmental "problems," and the seriousness with which the former are regarded is dependent upon the claims-making activities of the latter. Scientific theories about the environment tend to become "stories" which have social functions, but not explanatory power. This sociological reductionism ignores the diversity of theoretical positions in the natural sciences and the extent to which some of these have been critically aware of the effects of contemporary political opinions, cultural values and historical events in shaping scientific opinion (Pepper 1996:239–95).

Some strong constructionists are of the view that the "postmodern condition" may take us into a more environmentally benign "postindustrial" future with information technologies refiguring human relations with the environment. Donna Haraway famously argued that new technology and the social forms it generates, are refiguring the boundaries between humans, animals and machines (1991:165). We are becoming "cyborgs," hybrid embodiments (1991:150). Destabilization of boundaries illuminates the false "unitary" constructions of gender, race and class (1991:170–2), and by "queering" nature (Haraway 1994) we can acknowledge that the boundaries between the social and natural are "leaky." Strong constructionists argue that things in the world gain their character from social action (such as human knowledge, understanding and interpretation), rather than by virtue of any objective properties, and thus Haraway contends that "Nature cannot pre-exist its construction" (1992:296). Our knowledge of "things" does not have any objective basis, but is culturally relative; knowledge is dependent upon its "location" in time and space. Whilst some are happy to admit such conflationary of ontology and epistemology (van Loon 2000:170), this is rarely made explicit. An important question still remains: how might Haraway know that all our boundaries *are* leaking?

Because extreme social constructionism does not operationalize a separate ontology and epistemology, we cannot distinguish beings and objects in the world from our constructions of them. In an often-quoted example, Keith Tester asserts:

> A fish is only a fish if it is socially classified as one, and that classification is only concerned with fish to the extent that scaly things

analyses detach themselves from a deployment of discourse as symbolic regimes embedded in and constitutive of some kind of social/natural reality, and there ultimately seems little more in these analyses, than the stories we tell about "nature."

For both realists and social constructionists, "nature" is an ideological construct, and popular discourses of nature shift in content over time and cultural space. David Demeritt suggests that:

> If nature is socially constructed, its existence is not independent of our knowledge of it ... even if there were an ontologically independent real world, our empirical observation of it would still be biased by our socially constructed preconceptions of it. (2001:26)

This is where realists and constructionists part company. Demeritt is wrong – if nature is socially constructed it does not *follow* that its existence cannot be independent of our knowledge of it. In advocating a realist approach, Andrew Sayer argues that narrative approaches reify the context of knowledge so that no theorization of events, process and institutions can actually take place (1992:260–1). The non-human lifeworld exists in a concrete sense, and not only in human imaginings as narrative, and to assert the latter constitutes what Bhaskar (1979) calls "superidealism." "Nature" is a social construct, but that nature is constructed in terms of discourses, symbolic regimes that are *concretized* in institutions and processes. Importantly also, the environment has a reality beyond human ideas and beliefs.

Constituting the real

Realists argue that the environment consists of entities or beings with objective properties, with characteristics that are independent of social processes and human understandings. This means there is a "real" environment out there, which has certain properties and powers, whether humans understand those characteristics or not. The "critical" in critical realism emphasizes that we can both accept the idea that the world is composed of "real objects" with independent properties and causal powers, alongside an understanding of the social construction of that world in different ways by human subjects.

In *The Possibility of Naturalism* (1989), Bhaskar argued that although we can make similar kinds of observations about social and natural phenomena, we cannot analyze them in the same way because the structures that shape social life are so different from those shaping the natural world. Unlike natural structures, social structures are not independent of the activities they govern and do not exist independently of people's perceptions (1989:30–8). Benton (1985) considers that these differences are overstated, and that social structures can be independent of the activities they govern. For example, we may not be aware how social structures are shaping our activities and behavior. We can consider the ways in which social structures and natural structures interact, and Benton (1991) suggests, to this end, that certain approaches in biology might have important compatibilities with some approaches in sociology. This is precisely the view of Fritjof Capra who has used complexity theory in order to develop an approach to understanding both "social" and "natural" life in terms of systemic relations. Peter Dickens (1996:29, 2001:107) contends that the theories of life and ideas of the causal powers of natural objects as understood by biological and physical sciences, can be given additional depth and accuracy when combined with the understandings of contingent social factors.

For Dickens (1992), relations between any living organism and its environment need to be understood as mutually dependent and mutually constitutive (i.e, we cannot consider the "beingness" of any one living thing, without considering the environment with which it co-exists). Any living being has the potential to act, but the way in which they do so is shaped by the conditions in which the organism finds itself. This is a reciprocal process: the organism can have an impact on its environment, and the environment impacts on the organism. The problem with social constructionism is that it can only see this relationship as a one-way process where human thought and action constructs the environment. The realist position is that "things" have independent properties separable from human perceptions of them, so Dickens sees nature as having its own capacities, properties and powers. Human beings continually interact with and affect "nature" however, and there are contingent social circumstances that shape the environment. Whilst the environment has natural properties and powers, human practices and social behavior may shape, enable and constrain the processes of the natural world. For example, the human practices of farming, wildlife conservation and gardening all have implications for the various species of plants and animals for some are able to flourish and others are "constrained" from flourishing.

Benton (1994:31) argues that a critical realist approach facilitates greater analytical complexity because it does not transmute "nature" into symbolic representations alone – we can distinguish between the symbolic representation of the environment and the materiality of the environment. For constructionists, such a position is implausible, for we cannot get outside the symbolic order of cultural narratives on the environment in order to study the relation of the ideological representation of the environment to any concrete form "it" may assume. A critical and reflective realism is necessary in avoiding the relativist slippage of postmodern accounts that deconstruct the environment to the extent that any conception of human power over the non-human animate lifeworld is lost. Dominant ways of conceptualizing the environment have "real" effect, and concretize themselves in practices and institutions. What *is* essentialist, is a position in which the inevitability and prescience of social construction assumes social constructions are "true" (Fuss 1989:19–21). Even strict constructionists do not argue there are no "real" environmental issues, and as I argue in Chapter 4 with respect to feminist theory, strong constructionists make much of their radical contestation and contradiction of "the real," whilst almost inevitably reverting to some realism. In this sense, the battle between realism and constructionism is not so dichotomous, for strong constructionism is implausible.

Co-constructionism could be seen to put forward some form of "middle way," but it is problematically anthropocentric because it cannot allow for the independent properties and powers of natural systems. Bruno Latour (1993) argues that social and natural factors interrelate to such a degree, that it is meaningless to distinguish them. Latour (1987) suggests the natural and the social world are "co-constructed" by an intermingling of social and natural factors, positing an "amodern" ontology of hybridity, of not quite natural and not quite social objects. Thus what is natural and what is social is not easily separated out (Irwin 2001:162, Whatmore 1997). These hybrid conceptions of socialnature are chains of heterogeneous connections (Murdock 1997:745), which Latour calls "networks." Whereas Dickens wants to allow for the potential independence of the natural and the social, Latour suggests these categories are inseparable. Although Dickens' account is very different from Latours', they have similarities in considering the interrelations between the material (resources, technology, biology, economic production) and the ideological (culture, ideas, beliefs, representation). Latour (1993, 1999) can be seen as realist in arguing that humans are not the only actors who construct and change social and natural life, other species also "act." Social and "natural" entities constantly exchange properties within complex sets of relationships. I am not sure that Latour *et al.* do actually see objects, phenomena and processes as amalgams of the social and natural. The term "co-constructionism" almost by definition preserves the separation – if things are jointly constructed, they are "made" by two kinds of factors in interrelationship. By allowing the social and the biological to have the potential to be separate, we can account for a wider range of events and phenomena and examine the complex ways in which physical and social factors interact and interconnect. It is only by allowing the potential autonomy of natural processes that we can avoid seriously underestimating the emergent properties and causal powers of nature.

Much of the environment is characterized by the ontological hybridity, but we also need to allow for non-hybrid causality. Tidal waves and earthquakes as we have so painfully seen recently, are phenomena, which seriously impact upon human communities, but are in no way physically co-constructed. The migratory patterns of deer, whales and birds may be disrupted by human endeavor, but a disruption of pattern and process and hybridity are different things. Co-constructionist "actor-network theory" may respond with the useful notion of "actants" rather than actors, which are multiple and not species specific (Whatmore 1999:28, Murdock 1997:748). Yet I am still concerned with possible speciesism and anthropomorphism here. The kinds of actor-networks involving humans and those involving animals are of necessity understood differently. Whilst we can attribute intentionality and understand the social qualities of human action more readily than that of other animals, we need to allow the non-human relative autonomy from the social. This is the significance of critical realism – it builds in ontologically, the possibility of greater complexity by allowing a wider spectrum of difference. Perhaps Latour's (1998) greatest contribution is his insistence that the social sciences cannot *but* take account of "nature" because in all our spaces and places, humanity is embedded in relations with other species and actants. Whilst our lifeworld is messy however, it is ordered. The social power of humans becomes rather lost in the multiplicity of actor-networks, and there needs to be some conceptual apparatus which allows us to discuss the relative power of different kinds of actants. We are embedded in social and natural systems which constrain and enable our agency, and which we in turn shape and shift. These imbrications may be webs of reciprocal relations, networks of benign but unequal power, or involve domination and politically problematic practices of power.

Nature, society and complex systems

Environmental destruction is not necessarily a species generated problem as deep ecologists suggest, but one generated by particular groups of human beings operating in particular contexts – a product of both anthropocentrism and intra-human systems of oppression. Whilst social systems may be structured in part, according to the assumption that the environment exists only to serve human ends as theories of anthropocentrism suggest, humans *dominate* the environment, controlling, manipulating, exploiting and abusing. A more appropriate term, which suggests the complex patterning and extensive structures of human dominance, would be "anthroparchy" – human domination of nature. All ecologisms have failed to analyze a *social system* of oppression composed of structures, sets of oppressive relations, and I want to conceptualize human systemic domination of the environment as in articulation with other systems of intra-human domination. In order to argue for both systemic relations in nature and society and the intermeshing of such relations, I want to draw on some of the approaches to systems thinking which have emerged in the biological and physical sciences.

Complexity in living systems

Systems thinking, with its roots in biochemistry, has come to mean contextualizing and understanding relationships between phenomena, seeking to understand an integrated whole, the properties of which arise from the relations or connections between its parts. The properties of different elements of a system are not intrinsic, but can only be understood in the context of the system as a whole. Systems exist in a web of connections with other systems:

> Each of these forms a whole with respect to its parts while at the same time being a part of a larger whole. Thus, cells combine to form tissues, tissues to form organs, and organs to form organisms. These in turn exist within social systems and ecosystems. Throughout the living world, we find living systems nesting within other living systems. (Capra 1996:28)

At each level of structure and system, the term "emergent properties" has been used to describe specific qualities or properties which emerge at a certain level of systemic complexity, but which are not apparent at lower levels. Within scientific ecology, systems have become understood as communities of organisms which link together in a network fashion. In turn, networks interact with each other and operate within and across each other as "networks nesting within other networks" (1996:34–5).

What is most important in the understanding of natural systems is the form or "pattern" of the system, because system properties arise from configurations of ordered relationships. Living beings and systems are more than just atoms and molecules – they have a pattern of organization (Capra 1996:81), that is, the arrangement of components and their interrelationships that determine the essential characteristics of any given system. For Capra, the link between pattern and structure in living systems is the "process of life" – the activity involved in the "continual embodiment of the system's pattern of organization" (1996:155). Natural living systems can be said to posses a number of characteristics. First, they are "open systems" because they utilize a continual flux of and matter and energy in order to remain alive, and are dynamic and transformative. As such, the elements of the system are in a situation in which they must continually engage in self-regulation. This takes place by feedback, where, in systems of causally connected elements, the last element of the cycle feeds back into the first, and as it picks up systemic information along the way, it modifies its input (Capra 1996:51–62). There are a plethora of models developed of how natural systems regulate themselves.

One of the most influential theorists of self-organizing systems has been Ilya Prigogine who introduced the notion of "dissipative structures," or open systems. Through a series of experiments on chemical reactions it was found in situations far from equilibrium that coherent, structured, ordered behaviours emerged at critical points of instability (Prigogine and Stengers 1984:146), and can result in transformation – a jump to a new form of order and complexity. For Prigogine (1989), all living systems are both open and closed at the same time. They are structurally open to the flow of energy and matter, but organizationally closed, that is, despite the energy flowing through them, they maintain a stable form. The implications of this insight for social science are profound, and suggest a reconceptualization of the concepts of "structure" and "system."

The majority of contemporary social scientists have rejected the concepts of structure and system as reductionist in oversimplifying the complexity of social life and its multiple differences. The latter has become a key preoccupation (Felski 1997). The critique of systems theory has focused on an inability to account for the shifting nature of social life, and a preoccupation with notions of self-regulation or balancing in the maintenance of equilibrium, or social order, as apparent in the well-known work of Talcott Parsons (1960). Yet systems and structures in complexity thinking, are at once ordered yet disordered, stable yet unstable. As Prigogine (1980) surmises, we are compelled to consider the question not of "being" but of "becoming." Instabilities in the structures lead to new forms of order in living systems and these are often of ever increasing complexity, rather than, as Parsons suggested with respect to social systems, a return to a stable consensus after the disruption of pressure for social change. Whereas Parsons saw systems as moving gradually and consensually toward a Western model of modernization, Marx saw gradual change punctuated by revolutionary transformation along a predictable trajectory of class conflict. In complexity thinking however, systems and structures are not teleological – the path of systemic development depends on the systems history and various external conditions and cannot be predicted (Prigogine and Stengers 1984:140). In "nature" non-equilibrium is a source of order and what often appears chaotic exhibits complex patterns and is manifest in richness and diversity. Prigogine and Isabelle Stengers consider that this new theory of natural structures may lead us towards a new "dialogue" with the natural environment.

A second conception requiring some elaboration is that of the organization of life known as "autopoiesis." Neuroscientist Humberto Maturana (1980:xvii) advocated that the distinctive features of the organization of living systems were that they were "self-making." Controversially, Maturana postulated that, whether nor not organisms had nervous systems, in adapting and recreating the conditions of life, they should be seen as engaging in a process of cognition through making and remaking themselves and both constructing and delimiting their own boundaries. So, the process of knowing or cognition is identified with the process of "life" itself, and is immanent in matter at all levels of life from the very simple to the most complex. As Capra (1996:170) notes, this is revolutionary. Bacterium or plants in this model are seen as cognitive – they can perceive changes in their environment, and take appropriate action. Mind is not separate from matter, for cognition is the embodiment of action and responsiveness. Perhaps most importantly for ecologism – the environment has agency. Again, there is a challenge for established social systems theory – systems are both bounded and flexible – reproducing themselves and incorporating change.

As far as ecologisms are concerned, perhaps the most influential of these approaches has come from James Lovelock's "geophysics" or systems science of the earth. Through collaboration with the microbiologist Lynn Margulis, Lovelock developed the hypothesis that the earth was a system, a "superorganism" (Lovelock 2000:15) able to regulate its own temperature. Much of this regulation took place through the production and removal of certain gases from the earth's atmosphere by various organisms. A vast network of feedback loops brings about planetary self-regulation, and links together living and non-living systems. Complex cycles involving volcanic activity, rock weathering, bacteria and oceanic algae, all contribute to the regulation of atmospheric gases, climate, ocean salinity and other key planetary conditions. There is a tight web of interlocking relationships between the planet's living parts (animals, microorganisms and plants) and non-living parts (atmosphere, rocks, oceans). Living things, like animals, by obtaining energy from the sun or from food, "incessantly act to maintain their integrity" (2000:18). The earth system is also living by virtue of self-regulation and self-organization as emergent properties, in which open systems take in energy and remove waste.

Gaian evolutionary theory acknowledges the random genetic changes and natural selection so important for Darwinism, but its central organizing concept is the creativity of the evolutionary process whereby life evolves toward diversity and complexity (Kauffman 1993:173). Organisms do not merely adapt to their environment in order to survive and reproduce, the process of evolution involves the creation of novelty and of organisms that may or may not adapt to environmental conditions and may also alter those conditions. There is a complex interweaving of processes of co-operation, competition and co-adaptation, which characterize the evolution of life on earth. Margulis and Dorion Sagan have developed a notion of "symbiogenesis" – the creation of new life forms through the merging of different species (one living inside another for example) and argued that life should be conceptualized in terms of complex arrangements and developments of co-operation and creativity (1986:119). For Lovelock, the notion that the earth is alive, and is a single system within which multivariate networks of systems exist, leads to an understanding that "we are part of the Earth system and cannot survive without its sustenance" (2000:xx). Humans, and all other animal species are environmentally embedded:

> We live in a world that has been build by our ancestors, ancient and modern, and which is continually maintained by all things alive today. (2000:33)

The notion of a nesting of systems is also implicit in the notion of Gaia. Gaia is the mega system, and places a (planetary) boundary on all others which nest within it. Systems both diminish in scale and intensify in the proximity of their interrelations as we move from the planetary system, to ecosystems, to individual species and entities (2000:39). There is no hierarchy of "natural" systems however, they interlock and form a network, the totality of which constitutes the Gaian system. In this structural and systemic account therefore, the active processes of the constituent parts is crucial. The old social science "problem" of the relation of structure and agency is accounted for in the Gaian understanding of homeostasis: the active agents (multivariate organisms) both regulate various levels of system and are in turn implicated within and regulated by them.

As far as our understanding of contemporary environmental problems is concerned, Gaian theory offers some sobering understandings. Lovelock's mathematical modeling leads him to the conclusion that the impact, and Lovelock is specific on this (2000:154), of modern, rich, industrialized consumption orientated societies, on the composition of atmospheric gases is huge. It may be so disruptive, that the earth will shift radically to a new state where the climate is far less suitable for our species. Lovelock (1979:107) does not presume that feedback mechanisms will be able to overcome our impact on the earth's climate – rather, as Margulis (1995) has put it "Gaia is a tough bitch" and will eliminate (through self-regulatory mechanisms not an anthropomorphized teleology) those whose collective action makes life less comfortable. Human rights are not enough; Lovelock suggests "we must *also* take care of the earth" (2000:228, my emphasis). How these imperatives fit together is a key question for Capra, and for the arguments I pursue here.

In complexity science, the organization of living systems is a self-generating network, and Capra is keen to link such theory to political and philosophical understandings of deep ecology (1996:7). Drawing on the complexity and systems approaches that enable us to understand life as a web, or network, Capra suggests basic principles for ecology. These include: interdependence (the mutual dependence of all life process), holistic systems thinking and an understanding of the cyclical flow of resources, flexibility, diversity, cooperation and partnership. These all, he suggests, can and should apply to human communities and the environments in which they are embedded. Capra admits that deep ecology cannot tell us about "patterns of social organization that have brought about the current ecological crisis" (1996:8).What is most disappointing, is that he does not integrate the systemic theoretical insights from other ecologisms into his analysis in order to elaborate the specific properties of the social domination of nature.

Complexity in social systems

Important parallels might be drawn between "social" and "natural" systems in terms of complexity analytics. Social systems can be analyzed in terms of their emergent properties and powers, and social change is often both non-linear and unpredictable. In addition, social causation is complex – there may be multiple causes to be analyzed, and "the combined effect is not necessarily the sum of the separate effects" (Byrne 1998:14–15). For both complexity theorists and realist social scientists, the world we observe is the product of complex and contingent causal mechanisms, and thus the "real" may not always be clear and apparent to us (Byrne 1998:37–8). Complexity theory sees systems operating in hierarchies, and interactions between and within systems are not seen as additive, but as multiplicative, which sits well with the realist conception of complex and contingent causation. Complexity theory is a "scientific ontology" which fits well with the "philosophical ontology" of Bhaskarian realism because it:

> treat(s) nature and society as if they were ontologically open and historically constituted; hierarchically structured, yet interactively complex; non-reductive and indeterminate, yet amenable to rational explanation; capable of seeing nature as a "self-organizing" enterprise without succumbing to anthropomorphism. (Reed and Harvey 1992:359)

However, for Maturana and Francisco Varela (1987:89) an important difference between social and other living systems is the degree of independence individual system components have. For example, in an individual organism, the components of the system, cells, have a minimum degree of independence as they exist for the organism (1987:199). In human societies, the individual human components have a maximum degree of autonomy to act within systemic parameters and also to shift those boundaries. Due to these differences, Maturana (1988) and Varela (1980) in rather different ways have argued that social systems are unlikely to be "autopoietic" or self-making. In addition, given the agency of living compared to non-living systems, living systems demonstrate differing levels of unpredictability.

Within the social sciences, perhaps the best-known exponent of systems theory has argued that social systems *can* be autopoietic – as long as human social systems are defined as bounded by the domain of the social. Niklas Luhmann (1990) uses the example of a family as an autopoietic system with social boundaries and the feedback loops, which are sustained and constituted through conversations that establish shared contexts of meaning. Family roles and boundaries are continually maintained and negotiated (reproduced) through networks of intra-familial conversation. I return to Luhman in Chapter 4, but suffice to say here, in locating the human systems firmly within the parameter of the social, he extrapolates us from our embeddedness in ecosystems. When Luhmann speaks of the distinction between a social system and its "environment," he is referring not to natural systems, but to the useful complexity notion that every system takes all other systems as their context or environment and is interpenetrated by other system networks. As will be seen later in this book, I prefer the term "matrix" to describe the interlocking and interpenetrated complexity of social systems in particular.

In an attempt to account for such human embeddedness in both natural and social systems, Capra boldly proposes "A unified conceptual framework for the understanding of material and social structures" (2003:xv). His link between social theory and the complexity science of living systems is Maturana and Varela's understanding of cognition. He suggests two types of cognitive experience emerging at different levels of species complexity. These distinguish the "primary consciousness" of most mammals, some birds and other vertebrates which involves basic perception, sensory and emotional experience, and the "higher order" consciousness wherein the great apes possess self awareness, reflexive (social) consciousness and a sense of self (2003:34). Capra argues that many animals create their worlds by making distinctions and communicating them to others of their kind (2003:53). So rather than distinguishing "us" humans from other animals, reason places us on a continuum with them. Despite such linkages, behavioral rules, design, strategies and formations of power, are all but absent from the non-human world. It is only in human and chimpanzee social structures that design is manifested (2003:105). For Capra, understanding living systems in nature implies investigation of the relationships between form and process. When integrating the social realm we need to incorporate "meaning" – the plethora of characteristics associated with reflective consciousness. Capra is using rather different terminology here to discuss the thorny social scientific conundrum of structure and agency.

As we have seen, constructionist approaches conceptualize the environment as a culturally specific series of "narratives" having insignificant existence outside human consciousness. For realists, the world is differentiated and stratified, and composed of objects, including social structures, which have powers and capabilities (Archer 1996:694–6). Benton (1994) suggests that humans mediate relations with the environment in specific structural contexts, and that social relationships toward the environment should be thought of as specific sets of concrete social practices, which operate in a context of mutual dependence with the environment. Environment and society are partially autonomous, the environment can be theorized as belonging to the social, but it also exists as a series of "complex orders," which enable and constrain human activity (1994:49). Both the social and the natural have a "real" existence, and are characterized by structures, concretized sets of relationships and institutions (Sayer 1992:92). Neither humans nor non-human animals exist outside structures, which are a priori – we are born into a structured world, although agency reproduces and alters social structure (Bhaskar 1979:28–35). Structures have "emergent powers," irreducible to the individuals that live within them, and exist whether or not they are being exercised or suffered (Sayer 1992:119). Structure and agency are separate but interrelated phenomena, and just as the world has a reality separable from human experience, so do structures.

Anthony Giddens's famous "structuration" theory intended to place equal emphasis on social structure and human agency, by suggesting that they are interrelated to such a degree that they may not be properly distinguished. Structure is the mechanism through which action takes place and at the same time, agency produces and alters structure (Giddens 1979:5, 1988:288). Capra (2003:67) approves Giddens understanding, yet this is the right reading of Giddens *failure* to operationalize his own conflated conception of structure/agency due to its inherent difficulty. Margaret Archer (1995, 1996) has powerfully argued that conceptual conflation does not achieve its aim of linking structure and agency, but has the effect of "sinking one into the other" with the result that the interplays are lost. Rather, structure and agency are separate phenomena, which relate in various ways, and relations between them are spatially and temporally dynamic (1995:65). Giddens "compacting" denies that structure and agency both interrelate and exhibit independence (1996:688–9). Mouzelis (1997:116) points out there are historical variations in the structure/agency relation, yet Giddens presumes that they must be co-present (1979:67). In addition, the agency of which he speaks is human, and the structures it reflexively creates are intrahuman. Collier (1994:261) suggests such agential constructionism is thoroughly "anthropocentric." Society and nature are separate but interactive, interdependent, and both structured (Benton 1994:41). Differentiation within the animate non-human world is intense, and for certain species, there is a greater case for similarity than distinctiveness from humans. These multivariate differences need to be conceptualized in terms of structured relations of complexity. Once we have agreed that real entities exist in interrelation as part of ordered but highly complex structures within systems of relations, we can map the connections, divergences and disparities within/between/across systems and structures. Luhmann is wrong to consider this only within the social field. What is needed is an understanding of cultural and material relations, which allows, where relevant, that we may see the interplay of physicality and sociality.

Capra does not integrate social theories of system and structure into his model for the understanding of natural systems. Without an explicit inclusion of social structure, we loose the ability to properly integrate the ways structures constrain people's actions in relation to networks of power relations of domination. The non-conflationary model suggested by Archer, can also be seen as a complexity perspective – we can account for the unpredictability of social systems and structures, and their dynamic and transformatory properties. A particular collectivity may be subjected to the same structure, yet whilst some may act in ways consistent with the maintenance of the structure, others will act otherwise and in unpredictable ways. Considering the relationship between social systems and spaces, Byrne concludes that "the social" is both a system in and of itself, and the "space" within which a variety of systems are located. For example, households, systems in and of themselves, will be "nested" within the "condition space" of higher order systems (1998:25–8). Whilst nesting and symbiosis applies to the analysis of certain systemic interrelations, we also need a conceptual apparatus to explain the phenomenon of conflict within systems and between systems of similar and differing levels. Different systemic formations of social exclusion and social domination may have symbiotic relations at some historical junctures and in some places and spaces, and conflictual relations in others. The structures and processes within such systems may also be symbiotic and conflictual to differing degrees.

If we accept the notion of the planet as a "mega system," then this places physical and conceptual boundaries on our analysis. I am in agreement with Byrne that "society," however blurred its boundaries, might be analyzed as either or both, a system itself and one characterized by a range of systems nesting within it. Societies/cultures/communities are also embedded in locally and regionally diverse eco-systems of varying scales. As well as the flows of physical matter and energy, such ecosystems may be embedded in systemic relations of human domination. As Capra (1996:35) has noted, the ability to form hierarchies of group dominance and inclusion/exclusion is an exclusive quality of human social systems and absent from ecosystems or animal socialities, which are inclusive networks. Yet ecosystems exist within and across human systems of power and domination, and the structure of human social organization, involving the exploitation and abuse of the environment, implicates human communities, practices and institutions within ecological systems. Capra's (2003) attempt to engage social and scientific theory is impressive, but he over-compares social and natural systems and in so doing, looses the sense of social systemic relations toward the environment as currently structured around practices of exploitation and domination. Exclusive social systems of domination have long been a preoccupation of social theory. Gender, class, "race" for example have been analyzed as systemic formations of social domination. Nature is another construction implicated in webs of relations of dominatory power. The way in which the environment is structured requires analysis in terms of social and systemic properties in addition to the understanding of natural eco-systems. For example, in Western societies industrialization and the management of animal reproduction in agriculture, are key structures that shape the non-human animate world. Structural approaches to the environment are far more cautious than those considering intra-human stratification, and most retreat from an analysis of institutionalized human domination. Social systems and the multiplicity of ecosystems overlap, interlink, conflict and exist in conditions of symbiosis.

Anthroparchy: the social domination of nature

The "natural environment" is characterized by a ridiculous degree of difference. It is ridiculous, because a significant cultural premise of modernity has been to define "civilized" culture as above and beyond "nature." Such civilized culture is a notion that is interdefined by differences of class, "race," gender and other Otherings in addition to "nature." The symbolic separation of human "culture" from an amorphous "nature" or "natural environment" is a construction that is constituted by and through social institutions, processes and practices – through social structures. These institutions, processes and practices can be seen as sets of relations of power and domination which are consequential of normative practice and interrelate to form a network, a social system of natured domination or "anthroparchy."

Anthroparchy is a complex system of relationships in which the "environment" (i.e. living entities which are both themselves systems, and embedded in eco-systems) is dominated through formations of social organization, which privilege the human. Some deep ecologists have been mistaken in their presumption that we humans might effectively remove ourselves, or at least our deleterious impacts, from "natural" eco-systems, for we are embedded within them. Indeed, the global spread of human dwelling means that there is little left that might be approximate to an ideal of "wilderness." In Western modernity the "natural environment" has been conceptualized as a resource for human use and has for centuries been rendered hybrid by the interventions of human technologies. The globalizing tendencies of modernity (Giddens 1991) mean that relations of domination are not restricted to regions of the globe characterized by high modernity, but may be seen in different forms and degrees, in operation around the planet.

As a system of social domination, anthroparchy involves different degrees of formations and practices of power: oppression, exploitation and marginalization. I use these terms not only to capture the different degrees and levels at which social domination operates, but also the different formations it assumes within which some species and spaces may be implicated whilst others, inevitably, given their degree of difference, are not. Animals closer to humans in biology and sentiency can experience oppression. Different oppressive forms apply to different species due to their specific characteristics and normative behaviors such as the presence of sociality and the ways in which this presents itself. Exploitation has a broader applicability. It refers to the use of some being, space or entity as a resource for human ends, and one might speak of the exploitation of soils, woodland or particular plant species as well as the experiences of certain domesticated animals in agriculture, for example. Marginalization is most broadly applicable – the rendering of something as relatively insignificant. Whilst deep ecologists have been right to argue a case for "anthropocentrism," this can capture forms of human–environment relations that exemplify marginalization but is rather weak for the capture of more direct aspects of human domination.

Whilst the environment in all its infinite variety may be subject to anthroparchal relations, the agency of "nature" differs across time, space and context, and is implied by the ability of natural phenomena to exert their own properties and powers in specific situations. Tidal flows and a host of weather patterns may have considerable impacts on the ability of people to dominate their environments. Some parts of the lifeworld may not experience the effects of such dominations, whilst animals for example in intensive farming systems may keenly experience deprivation, frustration and fear. Some may feel that the term "human domination" is rather strong, but I do not mean to imply that all humans, in all places across time, are in dominatory relations to the environment, nor that all humans engage in exploitative and oppressive practices all of the time. "Humanity" is inevitably fractured by social and economic location and the interpenetration of cross cutting structures of various systems of domination mean that some groups of us are positioned in more potentially exploitative relations than others. In addition, individuals and collectivities choose not to exercise potential powers of domination and exclusion and also to contest them.

Systems and structures are dissipative; they contain the possibilities of dynamic shifts and restructurings. Social order is uncertain and impermanent and may also produce radical social transformations (Reed and Harvey 1996). Systems of social domination are inherently unstable, yet the complexity of their structuring builds in some form of relatively stable order. Some structures of anthroparchy may have intensified in degree whilst others may have lessened over time. These patters may differ when locality, region and/or globality are the frame of reference. Systems of social domination change over time and space but given the complexity of their networks and their links to other formations of domination, we cannot foresee what such changes might be, or predict a certain direction, though we might struggle to secure one. I suggest the following five structures network to form a social system we might call anthroparchy.

Anthroparchal relations in production

The production relations of anthroparchy have long involved the use of "nature" as a series of resources for the satisfaction of human ends. Production is a crucial link between humanity and "the environment" for as Dickens (1996) notes, following Marx, as a species, we interact with nature in order to survive by producing the things we need (such as food, fuel, resources for shelter). The mass production of goods and services associated with modernity in Europe and the industrialization of production significantly increased the ecological footprints of certain groups of humans, and the globalizing tendencies of Western industrial practices and process has led to industrialized production being a dominant structure shaping environment–society relations across much of the globe.

The technologies and institutions of industrial modes of production are tightly interwoven with the drive for profit maximization and the division of labor. Eco-socialism has been vital in illuminating the problematic and contradictory relationship between capitalism as a system of exploitative social relations and the commodification of nature as products, resources and waste. Whilst the global domination of capitalism is widespread however, the nature of production relations is shifting in certain important ways. Manuel Castells (1996,1997,1998) has convincingly argued that despite the diversification of society and culture across the globe, the penetration of capitalism as an economic system has never been so geographically expansive or so deeply penetrative of society and culture. Core economic operations are now global in scale, productivity is based around information and innovation, and the economy is structured mainly around flows of financial transactions (1996:434–5). Castells (2000) has a systemic approach to understanding the unfolding of this new formation of globalized capital, and has moved in the direction of complexity theory in arguing that the financial networks of international capital are inherently unstable. There is a sense in which this new economy is virtual and symbolic as speculation and unpredictable currency swings are key shapers of economic materiality, particularly of the fates of local and regional labor forces. For Castells, what is emerging is a fundamental contradiction between social cohesion and (diverse) cultural values and the logic of capital (1996:476, Boden 2000). A geography of exclusion affects vast regions of the globe, and is also embedded in parts of the more prosperous enclaves.

Structures of productive exclusion and the dominatory power of capital has led to the disruption of the society–environment spanning ecosystems, particularly of poorer regions of the globe. Economic globalization has resulted in a move of environmental "bads" from rich to poor regions, and a move of resources in the opposite direction. The neo-liberalism of the World Trade Organization, for example, has led to rapid depletion of natural resources (animal species, clean air and water) in the drive for poor countries to produce a few specialist goods for export in order to bring in foreign exchange. Environmental destruction and human exploitation is integral to this process, for as Yearley (1996) has shown, multinational corporations engage in "regulation flight." The productive imperatives of environmental domination have increased in ways that have impacted diversely on human communities in different regions of the globe (regulation flight has implications for the welfare of workers in "developing" countries and for unemployment levels in many parts of Europe, for example). Yet changes in production have resulted in successful contestation, as debates on resource and pollutant taxation attest.

Anthroparchal reproduction and domestication

Systemic domination shifts over time and new structures may emerge, or the relative significance of various structures shifts. One of the changing structures of anthroparchy involves the ability of humans to exercise domination over "nature" through the application of technology. Anthroparchal innovation has characterized human engagements with the environment for millennia, through the breeding (and hybridization) of plants for crops, and animals for food and labor. A matter on which all ecologisms agree is that the last two centuries have seen a problematic intensification of such processes through industrialized reproduction of plants and animals. Less directly, the reproductive systems of certain species are affected by human interventions such as commercial fishing, wherein the population structure of a species in a location shifts dramatically. As we will see, ecofeminists have noted the gendering of such reproductive interventions, particularly with respect to animal domestication and meat and other animal protein production. Such domestication may involve physical confinement, the appropriation of labor and fertility and entrapment. It may also operate at a symbolic level, for example, in the "need" to civilize and "tame" a wild nature, and the distinction between peoples, species and space, which are safely domesticated, and those dangerous beings and arenas that are not.

Most recently, all kinds of ecologism have become concerned with the development of biotechnology and, in particular, genetic transfer between species. Such biotechnology is often based on a determinism that reduces living organisms to passive collections of genes subject to environmental forces and constraints (Capra 2003:148–9). Because systems of domination operate in a network of relationships, synergies with capitalism are almost inevitable. Much genetic engineering is pursued by private companies who patent and sell their discoveries, whose primary interest is to secure profit by patenting transgenic plants and animals. Vandana Shiva (1993, 1998) has forcefully argued that the imposition of corporate monoculture has implications for the systemic relations between environment and society, particularly in "developing" countries. For example, patenting restrictions, enforced by the institutions of global capitalism such as the World Trade Organization ensure that traditional farming practices are disrupted, including organic methods of disease and pest control, and replaced by a biotech seed and pesticide package. Capra uses the language of anti-capitalism to describe this process as the commodification of life, but Shiva recognizes that the imperatives are not those of capital alone, but also of the exploitative context of post colonialism where the knowledges, practices, environments and communities of poor regions are disrupted and destroyed.

Through our increased capacity for technological intervention in "nature" and the increasingly "artificial" character of reproduction, this structure of human domination over "nature" has increased exponentially. Yet some of the most significant contestations of anthroparchy have revolved around questions of genetic modification, particularly of food crops and in Britain for example, the piloting of GMO crops has been withdrawn due to political agitation and consumer perception of risk. The European Union ban on battery cages for laying hens mediates some particularly harsh practice within animals' reproduction. Shifts in anthroparchal structures then, may be characterized by progress, regress and stasis at the same time, given the enormity of the relations of species and space they encompass, and all these complex changes need to be considered.

Anthroparchal politics

Institutions and practices of governance may reproduce, produce or contest and change relations of systemic domination. In much policy making, states at local, regional, national and intra national levels may be seen, almost by definition, to place the interests of the humans who run them, are subject to them or "citizens" of them, at the crux of decision making. States can act as direct or indirect agents of anthroparchy. Direct damage to ecosystems may involve, for example, subsidies for intensive farming, road-building schemes and destruction of ancient woodland. Less directly, states may encourage exploitation through apparent inaction such as not taxing resources. Yet states can also shift relations and practices of domination by the inclusion of certain kinds of rights to welfare and even self-determination for certain non-human animals, or placing other boundaries on human relations with the environment that limit our intervention in certain positive ways. Institutions and practices of political power reflect social relations of all our difference in systems of domination, and are shaped by prevalent discourses on nature and species, "race," gender and other exclusionary formations.

Violence

Violence is conceptually contestable. I take a broad sweep and include symbolic forms of violence, which may recall or suggest physical harm, as well as the most usual material definition of physical coercion. The definition of violence depends on both culturally specific and "real" notions of subjectivity, and normative presumptions shift over time, place and space. Whilst in Britain, domestic violence is no longer seen as legitimate chastisement of a man's chattel, we have witnessed fierce debates over whether chasing a fox and encouraging dogs to rip its body to fragments, is violence or amusement. Various ecologisms have debated the ethics of what humans might kill with impunity and what they should not, on grounds of a holist functionalism and an individualist sentiency. For animal species with greater levels of sentiency, violence can be seen to operate in similar ways to violences affecting humans. For example animals hunted, trapped, castrated or killed for food may experience pain and fear. Given that a key element of normative definitions of violence is physical damage, deep and feminist ecologies are right to include the destruction of habitats and eco-systems as a form of violence. However, whilst deep ecologists would contend such planetary violence is more problematic than the placing of hens in battery cages, some ecofeminists (Davis 1995) have argued that we might do better to "think like chickens" than like mountains, as a basis for an ethics of ecologism.

Cultures of exclusive humanism

Anthroparchal culture constructs notions of animality and humanity, culture and nature and other such dichotomies. It encourages high rates of consumption, it may represent nature in multiplicious ways that emphasizes the requisites of human domination, and suggest forms and practices this might take. Ecofeminists have had much to say on the interface between social nature and social gender, as seen in Chapter 5. Kay Anderson (2001:75) has powerfully drawn parallels between the constructions of "social nature" and "social race," contending that notions of culture and civilization have been constituted through an intermingling of gendered and natured discourse, thus culture is defined in terms of speciesism and ethnocentrism. I suggested at the start of this chapter that the human distinctiveness of sociality and the firm boundaries of "society" are based on the notion of human transcendence from and control over nature. Likewise, the notion of "proper" or civilized humanity has been developed in a provincial context (Soper 1995:66) of the centrism of European colonialism (Gregory 2001:87). The ideologies of "race" and "progress" associated with the latter, drew upon older, European notions of transcendence over "nature," particularly a movement from animality to "civilization." The interpenetration of social constructions of difference is a key site where various systems of social domination intersect. These constructions are linked to the ways in which social institutions operate and the lived materiality of those subject to the interpolated discourses of race, gender and nature.

None of these structures of anthroparchy are exclusively natured. They are cross cut, in different ways and to different degrees, by other formations of difference in systems of social domination. Capra suggests that the understanding of the interlinking between social and natural systems is part of a deep ecological perspective, but the complexity of his analysis provides a combination of systems ecology and eco-socialism. "Unfettered capitalism" (2003:185) he ultimately suggests, is the most significant threat to the environment and the key structuring mechanism of environment–society relations. This alone is insufficient.

Complexity theory has been important in reframing the concepts of system and structure and re-specifying the difficulty of the relation of social structure to human agency in ways that acknowledge our embeddedness amongst actants which are not human. Despite the enormous diversity and complexity of formations and relations of social and natural lifeworlds, we can consider the ways complex diversity is emergent in the behaviors of actants, prevailing discourses and discursive practices, structures and systems. Taking on board the simple but important notion that in a complex lifeworld, all systems take other systems as their context and thus interact with each other, I want to consider how a social system of anthroparchy might be interrelated with other systems of domination based on difference. This involves a consideration of how they might develop in ways that are mutually constitutive – as complex social systems are co-evolving. What we turn to now, are a range of understandings of gender relations, looking particularly at those that emphasize the systematic qualities and structural dynamics of such relations. We can then move towards an examination of how the systemic and structural complexities of "nature" intermesh with those of gender.

4
Different Feminisms

> it is grander being the earth, being nature, even being a cow, than being a cunt with no redeeming mythology.
>
> Andrea Dworkin (1983:184)

> We need the concept of patriarchy to understand ... (not only) the enduring problem of gender inequality, but the domination of our planet by individual and corporate masculine violence towards women, children, animals, nature and other men, ... But one of the problems with the term "patriarchy" is that it belongs to the very same compost heap which is home to the festering remains of gender and feminism. This is all rubbish, we cry, rinsing our hands with the anti-bacterial properties of postmodernism.
>
> Ann Oakley (2002:216)

The recent history of feminist theorizing has involved the marginalization of systemic accounts of gender, usually operationalizing a notion of "patriarchy" (and/or capitalism) in the understanding of social life. The "postmodern turn," has encouraged feminists to eschew systemic notions and emphasize the fluidity of gender as an identity constituted through human action. The cross cutting fractures caused by age, class, caste, religion, sexuality and spatial location, led some to conclude that the level of generalization required by systemic analyses made these forms of theorizing a homogenizing of social difference. This chapter argues against the grain, for a complex theory of patriarchy, which can account for the enmeshings of gender with/in/across other systemic formations of domination, based on "difference." Gregor McLellan (1995) has suggested that the theory of patriarchy advanced by Sylvia Walby

(1990, 1992) can be described as a form of "complex modernism." I argue this can also be seen as a form of complex systems theory, and a useful ontology for the understanding of gender in the context of cross cutting structures of other formations of social domination based on difference.

There has been little attempt to integrate an understanding of the naturing of gender into the mainstream of feminist accounts, and much feminist thinking can be seen as anthropocentric. Ecofeminism remains a minority formation and has been particularly prone to critiques of "essentialism." Whilst some radical feminists in the late 1970s and much of the 1980s demonstrated environmental concern, the notion of the social domination of nature was generally seen as a product of patriarchal ordering. I have already made the case both for a social system of natured domination and for the embedding of social systems in ecosystems, and do not think that a systemic theory of gender relations can have explanatory power across the matrix of difference-in-domination. It is more effective to consider systems of domination to be interlinked and intermeshed and also, as having specific histories, formations and constitutive elements.

Feminisms, systems theory and the problem of difference

The possibility of analyzing gender hierarchy as a "system" of social domination, characterized Western feminist debate from the beginning of the "second wave." In the 1970s, socialist and Marxist feminisms shared the assumption that women were systemically oppressed (Barrett and Phillips 1992:2), and the key areas of dispute around issues of domestic labor, reproduction and paid employment, situated such theorizing within the bounds of structural analysis. Socialist feminists argued that capitalism derived material benefits from women's socio-economic roles, and discussion focused on whether capitalism alone was responsible for gender oppression, or whether a "dual systems" approach was appropriate, in which patriarchy was seen as a separate system that interrelated with capitalism. From the late 1980s there was a shift toward poststructural and postmodern theorizing, wherein the certainties of anti-capitalist and anti-patriarchal theory have tended to be replaced by a fragmentary theorization of "difference" emphasizing the ideological representation of gender. Structural and systemic approaches to gender relations incurred strong criticism for exhibiting a "false universalism" (Eisenstein 1984) and an inability to account for difference between women. As a result, there has been a tendency for contemporary feminists to refrain from "grand theory" of explicative causation, and focus on micro-level analysis of localized and specific studies – a move from a structural analysis of "real" phenomena to one which emphasizes the symbolization of oppression (Barrett and Phillips 1992:4–7). Radical feminism however, continues to emphasize the importance of a systemic and structural analysis of patriarchy, a system of male domination of women. The institutions and processes that compose of patriarchy are conceptualized as webs of gendered relations, structures, which sustain and create formations of systemic social power.

Liberal feminist accounts like environmentalist perspectives, are reformist (Richards 1982). Early liberal feminists argued an ideology of gender difference articulated through popular culture restricts women's life chances (Friedan 1965). Naomi Wolf (1990) developed such analysis to account for changes in gender relations, arguing that the dominant gender myth of feminine domesticity has been supplanted by a "beauty myth" that disadvantages women. However, liberalism assumes that cultures of discrimination can be overcome by women's social mobility and renegotiations of their roles, facilitated by a pluralist state (Walter 1998). They emphasize an androgynous equality presumed on a fluid conception of gender, and here there are similarities to some postmodern approaches discussed in this chapter. Liberal feminists however, retain a liberal humanist and distinctly modernist analysis and agenda (Weedon 1987). In its uncritical articulation of Enlightenment humanism, liberalism can be seen as the most clearly nature-blind form of feminist theorizing, and its eschewing of concepts of structural and systemic social domination also make it of limited relevance in the development of a complex and systemic approach to gender and other dominations. All other feminist perspectives however, have contributed useful conceptual tools, or engage with debates that need to be addressed.

Capitalism and the systemic oppression of women

Socialist feminism has changed dramatically in the last quarter century, moving from revolutionary socialism to social democracy, a shift that has accompanied a demotion of concern with class, and an increasing tendency to focus on other kinds of "differences" between women. I would distinguish Marxist feminists as asserting the primacy of class in determining gender relations, which are analyzed as resultant from the operations of capitalism. Socialist feminism has sought to elucidate the complexities of gender and class whilst not granting primacy to either, sometimes adopting a "dualist approach."

Marxist feminism has generally seen gender inequality as derivative from the class oppression of capitalism. A "separate theory of gender relations," implied by the concept of patriarchy, continues to be seen by some as "confusing and unnecessary" (Pollert 1996:650). Class exploitation determines gender inequality that for many was rooted in the bourgeois family (McIntosh 1978), oppressing women through unpaid domestic labor (Dalla Costa 1973, Vogel 1983). Male workers are reproduced and maintained by women rendering the role of the housewife strategically important for capital (Malos 1980), yet this fails to explain why it is women who perform most domestic work, and are relegated to the "reserve army" of labor (Breugel 1979), or the ways in which household forms and relations differs according to ethnicity (Carby 1982, hooks 1982, 1984, Parmar 1984). Anna Pollert (1996) advocates the mutual constitution of gender and class, arguing that empirically, we cannot separate such concepts (1996:646). Not only does Pollert ignore ethnic difference, but also in arguing for the mutual constitution of gender and class, and advocating historical materialism as a means of theoretical explanation, gender is reduced to an effect of class relations.

Others have almost abandoned class in seeking to understand the politics of multiple differences. Iris Marion Young reflects a concern with "identity" with respect to a range of marginalized "groups" and notions of equality and "difference" (Young 1990:7). Young speaks of the "oppression" of certain groups through five "faces" of power: exploitation (of labor), marginalization (social exclusion), powerlessness (lack of political authority), cultural imperialism and violence (1990:56–64). If a group experiences any one of these five forms, they can be described as being oppressed. The degree of oppression varies depending on the extent of overlapping group membership, and the relation of such groups to institutions and practices of social domination. It is not clear whether Young sees these "faces" of oppression as structures or as some other kind of element in a system or plurality of systems of social domination. She does not explicitly deploy the concept of system, but uses the term "group oppression" in referring to sexism, racism, homophobia, agism and ableism (1990:132). Group analysis enables feminists to examine the similarities and differences across "cross-cutting" groups, enabling both an account of difference within oppressive relations, and a non-individuating politics based on an interactive radicalism of differentially oppressed groups. Harriet Bradley (1996) considers such accounts to be overly concerned with questions of identity and social fragmentation, and indeed this reification of difference does tend to mean a focus on a radical pluralist politics at the expense of an account of difference in terms of complex oppressive relations. Engagement with the problem of difference often means deploying "the same language of difference and multiple cleavage which dominate postmodern accounts," and some Marxist feminists assume this brings us "perilously close to the strategy of the deconstructionists" (Bottero 1998:485). Terminology alone however does not determine the analytics of a position and there are those who have usefully drawn elements of discursive analytics into materialist feminism (Hennessey 1993). This cannot be dismissed simply as postmodernist "slippage" (Gottfried 1998). Rather, there are attempts to integrate class analytics with gender and other differences in an analysis of structured social power.

The rich empirical work of Beverly Skeggs (1997) on the lived experiences of classed gender draws on Pierre Bourdieu's model of class as the distribution of various forms of capital. Skeggs develops Bourdieu's classification to include "embodied capital," in the form of "appearance" that working class women use to "trade up" in marriage and job markets (Skeggs 1997:10). Skeggs work reveals a complex picture of identities, which operates in a social space of gendered, classed, racialized and (hetero) sexualized "structures and power relations" (1997:160). For Skeggs, structural analysis is best retained within a single system conceptualized as "the social space," which is internally differentiated. If we differentiate, traditional Marxists assume that we fragment, and move away from structural analysis (Bottero 1998:486). Skeggs does conflate in arguing that gender is classed, and her theorization of class builds in more complexity than does her understanding of gender, and this suggests a need for conceptual distinction. Those who have accepted such a need for conceptual distinction between systemic structurings of gender and class have produced theorizations sometimes referred to as "dual systems" analyses, which consider two interrelating systems of capitalism and patriarchy. Zillah Eisenstein (1979, 1981) argues gender relations are produced through a single system of capitalist-patriarchy, yet attributes different structures to different systems. Heidi Hartmann (1979, 1981) avoids the problems of such conflation by conceptualizing the two systems as distinct but closely interacting. Women's labor is expropriated by capitalism and patriarchy in the form of domestic work and paid employment, and the former contributes to their disadvantage in the latter. These theorists acknowledge interrelations between different systems of social hierarchy, conceptualization of such systems as having some degree of autonomy and the identification of structures within which certain formations of exclusionary power relations cohere. However, Eisenstein and Hartmann over-emphasize symbiosis, failing to allow for potential disparity and conflict. In accounting for the complexity of social life, we must be open to the possibility of a multiplicity of systemic relations of inclusion/exclusion and/or domination, including of course, the social domination of nature.

"Black" feminism: the multiplicity of oppressions

A particular appeal of black feminism is the recognition of a "multiplicity of oppressions" (Smith 1995:694). Black feminist theory has been incredibly important in illuminating the ethnocentrism of some feminist theory and praxis, such as the homogenizing conceptualizations associated with radical feminist theories of patriarchy articulated in the 1970s and early 1980s (Davis 1981, hooks 1982, Hall 2002). Some have argued for a strategic centering of black women for feminist theorizing. If we theorize from the position of those most disadvantaged – "black working class women" (Davis 1990), the undermining of the oppressions that affect this most disadvantaged group will ensure "progressive change for all" (Davis 1990:31). Patricia Hill Collins (1989, 1990, 1995) has explicitly intermeshed this kind of argument with the epistemology of standpoint theory in order to suggest that black women occupy a particular social position in which they experience the convergence of an "inferior half of a series of dichotomies" (1990:70), for black women's experience of gender and "race" is often also bound up with class disadvantage. Race, gender and class are understood as dynamically interconnected in complex ways, with uncertain results (Crenshaw 1998), a notion of oppressive "multiplication" (King 1988), rather than addition where oppressive formations are interlocked and interdependent. Collins sees an afrocentric female epistemology as opening up feminist analysis to the conceptualization of individuals and groups as positioned within a matrix of oppressive relations in which they may be oppressed in some relations whilst privileged in others:

> Placing African-American women and other excluded groups in the center of analysis opens up possibilities for a both/and conceptual stance, one in which all groups possess varying amounts of power and privilege in *one historically created system*. (Collins 1990:225, my emphasis)

However, Collins's epistemology is at odds with her ontology and does not necessarily enable the complex account of differential power and privilege she suggests. There have been suspicions of any claim that

Different Feminisms

black women constitute a coherent group oppressed by white racis (Brah 1996), for black women's identity is highly fragmented (Aziz 1997), and any such "universalist" notion of black female identity is "essentialist" (Mirza 1997:5). The centering of African heritage is particularly strong in Collins's work, and is a problem. For example, it cannot properly account for the ways in which black American women might be privileged in relation to those in poor communities in poor countries.

Collins, like Skeggs earlier, however, is not focused on a politics and analytics of identity, but on the social forms oppressive relations take. Both consider a notion of a single oppressive "system" in which the multiplicities of oppression interconnect, but do not consider this in detail. A multiple systems approach would assist the exploration of different systems of domination in historically and spatially specific locales. The notion that oppression is interconnective has been a vital contribution, but in avoiding the systemic and structural approaches of socialist feminism, black feminism has often provided discussion of interconnections of a politics of resistance (hooks 1984, Smith 1995) rather than an analysis of the multiple connective oppressions to which it refers.

There is little work that attends to the naturing of race and of gender (Anderson 2001, Moeckli and Braun 2001) and the sedimentation of material exclusions, deprivations and marginalizations however, other than that undertaken within ecofeminism. The story of European culture, is a story also of the conquest of "nature," and those humans who have been natured are Othered by this process. Despite this, hooks (1982:112) deploys anthropocentric analogy in asserting that: "in sexist America ... black women have been labeled hamburger and white women prime rib." Ecofeminists have taken the difference of species seriously, and attempted to unpack its cross cutting with gender, "race" and other differences in domination. Feminists of all kinds have learnt not to use the metaphors of class, race and age domination in order to describe and conceptualize oppressive relations of gendered power. In order to properly open up feminism in the way Collins wishes, she and other black feminists must consider their anthropocentrism, just as ecofeminism must enhance its increasing inclusion of postcolonial perspectives.

Radical feminism: a patriarchal system of gender relations

Radical feminism has been distinctive in suggesting that both gender domination is systemic, and that it should be independently conceptualized, following Kate Millet's (1971) notion of a system of structured dominance in which men as a group dominate women as a group, and from which most men largely benefit. Critics have contended that the very term "patriarchy" "fixes gender relations in a transhistorical totality" (Gottfried 1998:455). However, it is not clear how the utterance of the "p" word implies that gender relations do not exhibit different forms and degrees of oppressive severity, across time and cultural space (Rowland and Klein 1996:14). Whilst some radical feminists use "patriarchy" descriptively, most refrain from any explicit deployment (Greer 1999). Notwithstanding, I consider that the understanding of patriarchy as a social system composed of webs of relationships that exhibit some degree of regularity is key to radical feminist thinking, and patriarchal relations are articulated in processes and institutions that form structures. Patriarchy is not emergent from other systemic dominations such as those based on class and race, but is autonomous.

Patriarchy is most often used to emphasize the social construction of gender as a system of social hierarchy (Mackinnon 1989) which is dynamic, not immutable, yet patriarchal theories have been accused of a homogenizing conception of men as "enemy" (Spelman 1990), which implies lesbian separatism (Segal 1987). There are incredibly few radical feminists who construe men as "enemies" (Gearheart 1982, Solonas 1983), and whilst "revolutionary" radical feminists (Coveney et al. 1984) advocated abstinence from heterosex as a strategy for patriarchal destabilization, this has not been a widely shared view, and does not involve "hating" men, just avoiding them. It is a simplistic reading of radical feminist theory that implies all men oppress all women, and to the same extent and in the same ways. Mackinnon (1994) acknowledges that men have unequal power in a patriarchal society, particularly if they are seen as "insufficiently masculine." Walby (1990) contends that to deploy a theory of "patriarchy" does not homogenize men, but allows us to distinguish "patriarchal men" from those who are not.

Theories of patriarchy have also been criticized as purely descriptive (Coward 1983), unable to explain the "origins" of male power (Bryson 2003). Some radical feminists have attempted to examine the origins of patriarchy and asserted that this lies with decline of matrilineal descent (Lerner 1986, Reed 1975) or Goddess worship (Stone 1977, Eisler 1990), or with the development of agriculture (Fisher 1979) or warfare (Starhawk 1990b). These theories do not claim to "prove" the origins of patriarchy, and as Millett points out: "Conjecture about origins is always frustrated by a lack of evidence" (1985:27-8). Such theories are best seen as contestationary stories that problematize stories about origins that stress the immutability of patriarchy, and in my view, the "origins" critique is misplaced.

Relatedly, patriarchal theory has been accused of false universalism (Coward 1983) and over-generalization (Lorde 1981, Ramazanoglu 1989). For example, Adrienne Rich (1977) has an account of patriarchy that seems to remove women from their social context in emphasizing a common experience of mothering. Black feminists have criticized such theorizing for obscuring racism (hooks 1982, Davis 1990) and argued it reflects a white, Western perspective (Lorde 1981). However, radical feminists have rarely argued that patriarchy is a historical constant or denied cross cutting influences of other oppressions (such as race: Griffin 1981, Bowen 1996, and class: Mahony and Zmroczek 1996). Segal (1987) is one of a long line of critics asserting that radical feminism sees women as passive victims. However, all radical feminisms emphasize the importance of political action (Dworkin 1988b, Bell and Klein 1996), and sees its theorizing as helping to identify political struggles (Spender 1985). Rather, radical feminists are often incredibly optimistic in the abilities of women, both individually and collectively; to change their everyday lived reality and wider social structures (Daly 2000). Some of the ways the theory of patriarchy has been used may exhibit a tendency to universalism that does not take account of the profound differences amongst women, and is insensitive to historical location, cultural norms and specificities (Dworkin 1974). However, this is neither inevitable nor endemic in the theory itself.

The institutions and processes that compose a patriarchal system are conceptualized as webs of gendered relations, which sustain and reproduce male social power. These structures include: law or the state, various sites of culture (such as religion, language, the media and education), the household, sexuality and reproduction and violence. Male violence against women constitutes a system of social control (Hanmer and Maynard 1987, Caputi 1988, Russell and Radford 1994). Rape and domestic battery has been understood as systemic and systematic (Mackinnon 1989:332) political acts which maintain certain power relations in which most men are privileged whether or not they carry out such acts of violence (Hanmer 1978:229). Others have seen pornography as a form of gendered violence assuming both physical and non-physical form (Mackinnon 1994, Dworkin 1981). Heterosexuality is a key institution of patriarchy organizing many aspects of gender relations (Johnson 1974, Rich 1980, Mackinnon 1989, Jeffreys 1990). Some have seen sexuality as a system of social stratification fused into one system with patriarchy, and speak of "hetero-patriarchy" (Hanmer 1989). Mackinnon sees gender, power and sexuality as very closely interacting, and at times, almost gives the impression that these categories are conflated (1989:126–31). She is trying to indicate the extensive nature of the sexualization of gendered relations of power and has more latterly adopted (Mackinnon 1994) a similar stance to Shiela Jeffreys (1990, 1994), arguing that patriarchy sexualizes inequality and that it is the "velvet glove on the iron fist" of gendered domination.

Others have emphasized the role of cultural institutions and forms in creating and reproducing male dominance, such as education (Spender 1980), language (Daly 1988), the media and popular culture (Caputi 1989, Spender 1995). For example, Spender (1980) has argued language is patriarchally controlled and a mechanism of enforcing subordination. She further argues (1983, 1995) that knowledge, be it academic, technological or popular is patriarchal and effectively obscures an understanding of male social power. Kappeler (1987) and Caputi (1989) have contended patriarchal ideology is carried by a variety of texts of popular culture, from romance novels to horror films. Daly (1973, 1979) has argued that sets of patriarchal ideas (religious, for example) are concretized in specific practices that are institutionally rooted. Those writing in the early 1970s (Greer 1970, Figes 1970, Morgan 1970) saw the household as a particularly important structure of gendered oppression, which sustains male power in the public world in addition to being itself oppressive due to domestic exploitation (Delphy 1980, 1984). Others argued it is sexual and reproductive expropriation in the household that is important (Firestone 1988:21). Structures of social domination interrelate however, and the forms and relations of household form and reproductive practices shape economic, legal and physical conditions of reproduction of both children, and domestic labor (Hartmann 1995).

One of the most widely read radical feminist works of the early second wave was Germaine Greer's *The Female Eunuch* (1970), which argued women were oppressed through a range of social structures including sexuality, relations in the household, employment opportunities and representation in popular culture. Greer has since been concerned with the ways in which other formations of domination cross cut with gender in terms of agism and its implications for women's relationships with their bodies (Greer 1991), and the questions of poverty, racism and underdevelopment in relation to reproductive politics in a postcolonial context (Greer 1985). In 1999 however, Greer published *The Whole Woman* in reaction to what she saw as the political, cultural and academic complacency about the social position of women, returning to the specificity of gender as a formation *despite* women's difference (1999:1–2). Unlike Caroline Ramazanoglu (1989) who suggested "women" are a contradiction in their fractured identities of difference, Greer argues that the ubiquity of sexual difference means feminism should take account of a plethora of dominatory relations. Her thesis is that Western women who are relatively privileged have bought the spurious version of equality offered to them by liberal democratic states, whilst the marginalization and exploitation of women in poor countries means that feminist arguments have to no extent been addressed (1999:330). "Ex-feminists" (Coward 1999, Hakim 1995) and anti-feminists (Philips 1999), have argued that in the twenty-first century women collectively choose the niche they occupy in employment and in domestic life. Greer argues however, that across the different sites of patriarchal structures in Western locales, we have seen progress, regress and stasis, and these differences need accounting for. The levels of acceptable violence in popular culture are an indication that this is less feminist than 30 years ago (Greer 1999:157). British women may appear to be making inroads to the job market, yet this is not what it was in terms of working hours and rates of pay, and evidence of fraternalism undermines liberal arguments for gender equality (1999:165). Men still undertake an insignificant proportion of childcare and domestic labor, and studies indicate that women are undertaking more housework hours than ever before, using more chemical products to do so, and consuming at environmentally unsustainable levels (1999:129–44). The "flight from fatherhood" thesis (1999:211) is empirically substantiated, she argues and we live in a gender-segregated society in which men engage in particular cultural pursuits. The burgeoning beauty industry is a nexus of capitalist patriarchal values and practices designed to reproduce and reconstitute the white Western cultural disgust of a women's body, which is exported through corporate advertising (1999:27–32).

The analytic strength of radical feminism is that it enables us to argue against such patriarchal social structures, which are *en*acted by men and women as social agents and reproducers. Without a systemic and structural account of social life, we lose feminism to positivistic arguments promoted by the mass media, or to the individualism of postmodern analytics. To argue that patriarchy is an autonomous and structured system of oppression is not to exclude the possibility that it interacts with other oppressive forms such as class and race, or nature, and a challenge for the feminisms of the twenty-first century is to come to grips with such complexity of difference in relations of domination.

Deconstructing domination

Feminists influenced by postmodernism have abandoned any attempt to capture the complexity of social relations through analyzing structural constraints, for approaches that stress the fluidity of meaning and the temporality of gender. Radical feminist approaches, particularly older texts (Richardson 1996:143) are held to posit an "essentialist" authentic female experience, which obscures differences created by cross cutting oppressions (Spelman 1988, Frazer and Nicholson 1990:2). Radical feminism is caricatured by such critique, with a few theorists repeatedly cited (Alcoff 1988:408–14). Mackinnon (1996:50–2) contends postmodernist critiques of patriarchy make the inconsistent assumption that the concepts of race and class are "real," in deconstructing gender, which is not, and are themselves reductionist in homogenizing "white women" as not oppressed to any degree (1996:52). Gregor McLellan (1995:404) further argues that the positions of some postmodern feminists, is not "feminist," for the "uniform features of gender identity (are) definitively subsidiary to other differences." This said, many of the best-known theorists retreat from much of their postmodernism in demonstrating an allegiance to some standard via which to evaluate theory. Jane Flax for example, is concerned that postmodernism may commit the "fallacy" of "presuppositionlessness" (1990:224). Linda Nicholson compromises her postmodernism by re-introducing structure and even system into discursive analysis: "by admitting *big categories* into narrative accounts, we can acknowledge the possibility of *structural features of societies remaining relatively static* over time" (1992:98, my emphasis). The work of Haraway (1997) is particularly inconsistent, deploying modernist categories to critique for example, the "cyborg" representation of humans as sexist and racist (1997:225–65). Thus much postmodern feminism may not be quite so "post" as it may at first appear.

The reification of difference

"Essentialism" has proved a most popular and often carelessly used criticism made of theories that are held to reflect "a belief in the real, true essence of phenomena" (Fuss 1989:xi). Those whose theories and concepts are judged to be "essentialist" apparently express, implicitly or explicitly, the idea that "things" have fixed properties throughout time. Postmodern feminists have rejected the categories of men and women as prone to naturalization, or at least, to the "social" essentialism of assuming static social divisions between men and women (Ferguson 1989:54). "Women," "men" and "gender" lack conceptual coherence and are culturally and historically variable, they are misleading labels that obscure the diverse realities they claim to represent (Flax 1986, Nicholson 1990), and Jane Flax goes as far to suggest that any process of conceptualization per se is irrelevant (Flax 1992:457).

Yet critiques that deploy "essentialism" rely on the same framework they purport to reject, "a master narrative of truth" which judges a theory "to be false from a position which is outside all positions" (Thompson 1996:334, Fuss 1989:2–6). It is not clear that to conceptualize gender denies plurality, or change (Weedon 1987:105). Somer Brodribb (1992) has strongly argued that postmodern deconstructionism denies both the reality of gender as a web of material practices embedded in social institutions, and the corporeality of women's bodies. Postmodern feminist theorists of the body (Grosz 1994) reflect their origins in literary and cultural studies, conceptualizing physical bodies as "texts" on which anything may be "inscribed" (Klein 1996:350), and failing to capture dynamics of gendered power embodied in certain physical practices (Spretnak 1996:323). In suggesting, for example, that the body be seen as "the medium" which must be "destroyed" and "transfigured" for "culture to emerge" (Butler 1990:130), we ironically return to a key Enlightenment motif – a privileging of mind over matter and a fully disembodied anthropocentrism.

Some have been concerned that in deconstructing women and gender, feminist politics becomes impossible, and Sabina Lovibond has argued this may be "politically convenient" for those advantaged by the project of modernity (1989:22; Di Stephano 1990, Waters 1996). Mary Maynard, defending the concepts "race" and "black," argues such generalized categories are of importance to a politics of resistance, and their deconstruction is linked to a "benign pluralism" (1994:11). Some black feminists have voiced concern that conceptual deconstruction may displace the understanding of racism and gender oppression obtained via an Afro-centric feminist epistemology (hooks 1991, Collins 1990). Feminists attracted to postmodern approaches often still see a political need for concepts of race, class and gender (Phillips 1992:28), and even Butler in one of her weaker moments concedes "there is some political necessity to speak as and for women" (1993:15). Walby (1992) argues that the concepts of gender, race and class should be retained on analytic merit, and postmodern approaches have gone "too far" in their emphasis on fragmentation, for in their dispersal of notions of power and identity, they ignore social context and "preclude the possibility of noting the extent to which one social group is oppressed by another" (1992:35). Or indeed, they preclude the possibility of noting very much at all (Bartky 1990:8).

"Oppression" and "difference" are not concepts inhabiting exclusive terrains policed respectively by modernist and postmodernist frames. Both positivists and postmodernists have associated difference with a transparency of identity. The position that one can only be "oppressed" if one knows it, is untenable. Oppressed groups should not have privileged epistemological status, for on top of the complex interweaving of multiple inclusions and exclusions, there is always an agential, individual self who articulates an "identity." If we loosen the association between difference and identity then we come to a position where the material and imagined differences of age and generation, sex, sexuality and gender, "race" and ethnicity, class, caste, wealth and other distinction can be incorporated into a complexity of systemic and structural Othering. The critical realist position enables us to have real-and-contested difference immanent in networks of relations of social domination. It also enables consideration of difference beyond our "species" and contestation of species as a political category. Now this would be a real challenge to the "grand narratives" of modernity.

(En)acting the narrative – gender as performance

Butler's influential writings articulate perhaps one of the mostly clearly postmodern of feminisms, and are distinctive in that reference to social structure is almost entirely absent. Her better-known earlier work focuses on the fragmentation of identity, specifically whether there is any coherence to the category "lesbian" (Butler 1990:5), but this is intertwined with the question of whether "woman" can be regarded as any kind of unified subject.

Butler's position is that identity is socially constructed via action, and gender, or any form of social structure is not a priori, for "the 'doer' is invariably constructed through the deed" (1990:142). The self constructs the acts and is thereby constructed through "sustained social performances" (1990:141). We are compelled to act within the gender identities of the historical and cultural location in which we find ourselves and we "repeat" behaviors (1990:140), such as gestures and movements, in terms of gender identities and relations. Change comes from the internal disruption of such gender categories (Butler 1993:91) primarily via parody in the form of drag which: "implicitly reveals the imitative structure of gender itself – as well as its contingency" (1990:137), because it does not parody authentic femininity, but the concept of such authenticity. However, it is a reductionist position which contends that gender is exclusively constructed by our action, and it is unclear how gender "performance" is to be changed unless women can "have a say in the production of the play" (Benhabib 1992:215). Agents are not necessarily aware of dominatory power relations (as Butler later acknowledges 1997), and as Jeffreys (1994) notes, it is uncertain whether the majority of women realize that a whole new range of gender identities is actually "open" to them.

Butler reduces gender to subjectivity. It is possible to recognize "women" as culturally differentiated and discursively contingent whilst also recognizing "women" as "real," and as subjected to various formations of gendered exclusions inevitably intermeshed with and lived through other categories such as class, age and "race." The closest to an admission of social structure in Butler, is the notion of the "heterosexual matrix" which is "an epistemic regime of presumptive heterosexuality" (1990:x), that is "oppositionally and hierarchically defined through the compulsory practice of heterosexuality" (1990:151), and could be interpreted to mean a set of institutions, practices, and ideas which shape/constrain human agents. However, because Butler (1993:xi) explicitly renounces externality, that is, any "thing" that produces a construct, there is no material "real." Thus Butler can write a book "about" the body without speaking much of bodies at all, and this flight from corporeality is rightly critiqued as "intellectual anorexia" (Zita 1992:126). Some argue that Butler secures a more "embodied" way of thinking (Nash 2000:654) through the performance of sedimented social practices. Yet Butler's subject is an abstraction from lived experience and embeddedness in time, space and place (Nelson 1999). I find it difficult to see how a conception of gender relations as "fabrications manufactured and sustained" through "signs" (1990:136) can apply to material practices of domination such as domestic violence or segregation in the workplace.

Butler's position here has similarities with Giddens's theory of structuration (1979, 1984). Crucial to structuration is a conception of the active agent, and people must be able to choose to reproduce social institutions or alter them in their reproduction of social relations (1979:106), and I think Butler operationalizes Giddens's sense of agency as constitutive of social life and reproduced in daily practices. A difficulty with the Giddens approach is his over-emphasis on the "minutiae of everyday activities" (Archer 1996:688) for social structure is something more than "practices" which, when regularized become "institutions." Social structures, sets of institutional, organizational and procedural relationships have a degree of continuity and regularity, and can be said to pre-exist successive cohorts of agents, whilst also being dependent on agents for their replication and alteration (1996:696–7). Archer (1995) argues structural properties are often resistant to change, and have properties that can be ontologically established and cannot be reduced to their constituents (Sayer 1992:119). Structures are real objects with emergent properties that have powers and liabilities, existing regardless of our interpretation of them. This critique is apposite for Butler, for the emphasis on human agency as constitutive of social relations, exemplified by a theory of performativity, precludes a conception of social relations as involving dominatory power. Whilst gender relations are dynamic, they exhibit regularity and continuity, and have a real existence beyond our "performance" of them.

There are layers of institutions and related practices (structures) within which performance is embedded. Cross cutting fracturings of social difference, along with the individuation of personality impinge on the abstracted subject, just as structural conditions are implicated in any human action. If we are to capture the embedding of subjects in specific historical moments and places, in complex patterns of relationships in which the matrix of difference-in-domination means they are experientially fractured in advantage and disadvantage, then we need an understanding of that matrix, and its shifting spatial and historical constitutions. We require a complex understanding of structure within which agency is embedded. The reification of the abstract agent, prey to the ever fracturing of their identities is a path back to a pluralism of a very old fashioned sort.

Patriarchy and the complexities of domination

As we have seen, radical, Marxist and socialist feminism have deployed both implicit and explicit understandings of system and structure in seeking to understand the social domination of gender. These positions have been charged with an inability to capture what black feminists have referred to as the "multiplicity" of oppression. A way of accounting for such multiplicity is to draw on the conceptual developments of system and structure emerging from complexity theory.

Complexity theorists have enhanced the detail and specificity of systems theory by elaborating layers of ontological depth. Niklas Luhmann's notion of social system is of self-regulating systems, which are autonomous from their "environment," that is, the surrounding milieu of the system – that which is beyond the system boundaries. Luhmann's systems are differentiated into autonomous subsystems and subsubsystems, each with its own distinct level of organized complexity. Thus, system elements are themselves systems and form the internal environments of social systems, and systems are nested, embedded within other systems, producing a hierarchy of levels of social structure. Louis Althusser (1963, 1968) suggested that the problem of teleology might be avoided if the levels of social structure are conceptualized as having "relative autonomy" within a "complexly articulated whole." Levels of social structure interconnect with and influence each other, but are not clearly determined by a single logic. Sylvia Walby (1990) uses this notion of relative autonomy to stress both the interlinked yet semi-dependent quality of the structures constitutive of her model of a system of patriarchy, and also, to describe the relationships between different systems of social domination. Walby does not embody complexity theory in her writings of the 1990s, but the basis is there for her more recent move in this direction in seeking to capture the complex structuring and restructuring of multiple social systems.

A complex theory of patriarchy

Walby has argued for an approach to gendered social relations in which structures are "constantly recreated and changed by the social actions of which they are composed" (1997:7). Social structures are "realities in flux and motion" (Godelier 1984:18), they continually alter, shift and re/form through human agency, an understanding commensurate with complexity analytics. Walby exemplifies the approach suggested by Archer (1995, 1996), in that her structures are real, have concrete effect, are composed of closely interacting sets of institutions, roles and practices that exhibit continuity over time (they are "relatively enduring," Collier 1994:16) and demonstrate certain regularities. They are structures of power relations that involve oppression and exploitation, but they differ across time and cultural, national, regional and local space (Walby 1997:7–12), and they intermesh to form a social system of patriarchy.

Walby (1990) is careful to specify that her model of patriarchy is developed in the context of liberal democratic late modernity and does not make any claim to universality. This is a contrast to earlier approaches, which either focused on one structure (such as kinship, Delphy 1977) or posited a general normative pattern organizing social life (Millet 1971, Hartsock 1984) and were open to the critique of Eurocentrism for the assumption of cultural homogeneity (Said 1978). Latterly, Walby has become particularly interested in the spatial divergence of complex social systems. She has contended that the processes of globalization have resulted in different impacts for social systems based on class, gender and ethnic structuring (Walby 2003b:5–8). For example, globalization has been associated with increases in both class inequalities and those associated with nationally based ethnically associated inequities within Northern regions of the globe. The change in gender inequality however, is more complex, for in certain locations, globalization has resulted in a decrease in gender inequality and a "modernization" of patriarchal relations.

The concept of patriarchy can grasp both institutional and non-institutional aspects of women's oppression, and Walby speaks of patriarchal relations in social institutions and in relations of normative praxis. Her use of conceptions of relational structure is linked to her deployment of systems theory, and the notion that systems and structures are cross cut by various kinds of social difference. Patriarchy is conceptualized as a system of social structures in which men dominate, exploit and oppress women. Patriarchal structures are important sets of relations of power, which are deep-seated, and not always readily apparent, having "ontological depth" (Bhaskar 1978). This is a "strong" critical realism in which structures are conceived of as transphenomenal (going beyond appearances), and counter phenomenal (sometimes contradicting appearances) (Collier 1994:6–7). Whilst Walby sees structures as limiting, they are not determining (as critics suggest, Pollert 1996:639), and whilst human agency involves "constrained opportunities" (1997:7), patriarchy changes (restructures) through feminist contestation. Bhaskar (1979:35) similarly argues structure reproduces and changes via human action, with the reproduction of structures the most common form of human action.

More recently, Walby is particularly concerned to show the transformative quality of gendered structures. In Britain for example, she contends that we can see gender convergence with men amongst more privileged women (young, educated, employed), shifting formal political relations with the increase of state feminism, and also entrenched relations of patriarchal inequality in other areas and involving groups of women differentially stratified (Walby 1997:22–66). Actors recreate social structures, individually and collectively, in ways that reinforce and alter relations of gender. She has drawn on complexity theory in developing the notion of "waves" of repercussive events, which build through "endogenous processes" into social movements which spread out from one spatial location affecting social relations in other locations (Walby 2003b:16). Such waves carry social and political projects such as those of ecologism, feminism, anti-capitalism and human rights, although they are not necessarily associated with progressive visions of the social order. Waves are sources of "social energy" (2003b:17), forces which pass across and through institutions and patterns of social relations, and may exhibit its force and effect differentially over time, for example, through waves of feminist activity. Walby's (1990) model of a patriarchal system assumed adaptation to a dynamic environment. As a result of both feminist contestation and shifting interrelations between patriarchy and other systems of social exclusion, Britain has seen a historical shift in patriarchal form from private to public. Women are no longer controlled individually by men within the household and are no longer excluded from both power and public, but are controlled collectively, primarily through the state and paid employment. Contemporary patriarchy does not exclude women but maintains control by segregating them in subordinate roles, a lessening in some areas of the oppressive degree of patriarchal relations. The context of anti-patriarchal contestation is also shifting, with feminism increasingly articulated through a global discourse on human rights, utilized in different ways by regional movements, in the context of differing levels of economic development (2003b:20). The acknowledgment of patriarchy as a system that alters in specific ways across time avoids the criticism of ahistoricism (Rowbotham 1979) leveled at earlier accounts. Walby provides a wide-ranging yet complex and historically and culturally specific account that examines interpenetration of systemic domination in terms of both symbiosis and tension, using the notion of the "restructuring" of gender relations in order to capture the spatial unevenness of social and political change.

It is notable however, that since the mid-1990s, Walby has abandoned the "p" word, for the use of "gender regime." She sees this as a "translation" of patriarchy into contemporary terminology, so a regime is used as a direct alternative to the controversial "system." A "regime" connotes "the systematic relations between elements of a system" (Walby 2004:8) which may involve a range of social institutions and their associated practices at any particular time. Walby has moved more firmly into multiple systems theory here as she presumes that a range of different kinds of regime will co-exist. In any particular kind of regime, different forms may be present. She retains the public or private notion of gender regime/patriarchy, but further differentiates this into market-led, welfare state-led and regulatory/polity-led paths of regime development (Walby 2004:10–11), in a manner similar to those who have spoken of "varieties" of forms of capitalism (Hall and Soskice 2001). Here, the restructuring of the gender regime involves different routes, depending on whether the state or the market is involved in the provision of certain domestic labor (such as child care), and these routes or pathways produce different kinds and degrees of social transformation. I am not convinced of the need to change terminology, and consider that "patriarchy" better conveys a broad spectrum of relations of social domination than the governance implied by the notion of regime. We might have varieties of patriarchy to capture specific historical and cultural formations, but I consider that the power of "patriarchy" to evoke systemic gendered oppression is pertinent and worth retaining, despite the current knee-jerk reaction in academe.

A criticism I would make is that more complex theories of patriarchy say little of what Scott and Lopez (2000:4) call "embodied structure." This concerns "the habits and skills that are inscribed in human bodies and minds that allow them to produce, reproduce and transform institutional structures and relational structures." The ontological depth of social structures and social systems is inscribed into human (and other) bodies. Radical and ecological feminisms have emphasized this in particular, and it is also crucial for ecological theories of system and structure. Chapter 6 will consider the embodiment of institutional and relational structure as played out specifically on bodies of women and some animals.

In countering a political culture of post feminism and an intellectual climate of postmodernism, some feminists are returning to an analytics of system and structure in order to understand the complexity and fracturing of gender and other relations of social domination without resorting to an individualist theorizing suggested by postmodern deconstruction (Oakley 2002:27). Walby's account of patriarchy can capture the difference of gendered domination because it implicitly builds in elements of a complexity approach.

Complexity in social system and structure

In the social sciences, Niklas Luhmann has made the most concerted attempt to work through the implications of developments in complexity science. Luhmann does not present a specific theory of society, but provides a toolbox, a set of conceptual instruments for deployment by those of us who might so wish to develop such a theory. Luhmann (1995) understands society as a complex system of communications, differentiated horizontally into a network of interconnected subsystems. Following Humberto Maturana and Francisco Varela (1987), Luhmann uses "autopoiesis" to suggest that each system reproduces itself on the basis of its own internal operations and system specific attributes, has a form of consciousness and observes its specific "environment" (the milieu outside the system boundaries) from which it differentiates itself.

What makes Luhmann's work of particular interest for my own project is that it begins and ends with difference, rather than being preoccupied with synthesis. Luhmann's concept of difference incorporated within systems as "differentiation" means that each system is "itself a multiplicity of system/environment differences" (1995:18). Systems differentiate by selection, and this leads to the formation of systems that are less complex than the milieu ("environment") in which they exist. Luhmann turns agential theory on its own sword, in arguing that such approaches are incapable of dealing with the multiplicity of social difference because they conflate difference with identity and selfhood. For Luhmann, difference and identity is not the same thing; rather, "identities" are "introduced to organize differences" (1995:75). The notion of identity cannot properly capture the elements of difference that are not necessarily actualized in individual or collective subjectivity. The very difficulty with Butler (1993) for example, is that she begins, as she says, "with the 'I'," the subjective self. This is inevitably the reified and generalized self, for the infinite difference of individual human agents actions compel agential theory to create a highly abstracted transcendental subject as the basis of knowledge and claims making. Human action for Luhmann (1995:164–75) always occurs in situations, and thus the actions are not the ultimate ontological given that action theorists assume, rather, social systems are also "action systems." What we end up with if we traverse the agential path, Luhmann cautions, is at worst, the methodological individualism of rational choice theory, at best, symbolic interactionism (1995:108, 256). Butler's "agential realism" results in the latter.

In Luhmann, the non-system environment for that system is that complex mass of phenomena and processes from which specific systems differentiate themselves (1995:177–81). Society is itself a social system, inevitably of greater complexity than the multiplicity of systems that nest within it, and again, is less complex than the range of non-social systems that in turn, form its own environment (1995:183). This process of differentiation internally structures social systems in terms of sets of relations and levels of system operation, and system building is bound up with the process of selection on the basis of difference (1995:134). In the latter, we may have social systems "nested" within more complex social systems. In the former, I think Luhmann is trying to capture a notion of social structure. We cannot include "people" as the parts of a social system but require middle level concepts, as Luhmann puts it, chapters not letters, form parts of a book, and rooms, not bricks parts of a house. I would say that within social system(s), structures are less abstracted orderings of the social than the conception of system. The relations within and between the internal elements (structures) of a system constitute a form of regulation ("conditioning") of the system as a whole (1995:23). This takes the form for example, of rules of social inclusion and exclusion, which establish "conditions of possibility" within which actants may act and structures "restructure" themselves. Systems are continually adapting to "internal improbabilities and inadequacies" (1995:31) – they are utterly dynamic. Restructuring is inevitable because regimes of human meaning are autopoetic – the articulation of meaning involves the emergence of new differences that will be incorporated, thereby changing the system (1995:137–45). Communication is the means of structuring and restructuring of social systems, by which boundaries shift and are policed, and new/old elements are included and excluded. Thus the reproduction of social systems cannot simply be seen as the "replication of the same," but rather as a constantly "new constitution of events" (1995:189).

Luhmann uses Parsons concept of "interpenetration" to characterize the interdependencies between systems that result from the complex co-evolution of systems. Adapting Maturana's notion of "structural coupling," Luhmann (1995:220) argues that different systems rely on each other's complexity to further elaborate on their own internal complexity. Interpenetration is a complex processes involving selection. As the structural forms of social systems are often very different, interpenetration involves the cross cutting of certain elements and structures of particular systems and not others (1995:213–4). Luhmann argues that systems may "couple" temporally, for a specific period in a particular arena, occasion, or for a particular systemic purpose/logic, systems may operate as a unified entity (1995:223). This is an important notion for multiple systems theory – in specific arenas, systems of social domination may be structurally fused, and this is a dynamic and bounded event.

Despite an analysis that embeds the dynamic qualities of structure on every page, Luhmann's view of the capacity of systems to change through collective action is depressingly static. The enhancement of complexity through interpenetration is ultimately stabilizing and consolidating (1995:218, 284). The uncertainty of events and outcomes is key to the dynamic qualities of social systems, yet for Luhmann, social structures seek to maximize their system flexibility by ensuring that agents will endure maximum levels of insecurity (1995:320–3). Social structures are also structures of expectation, and the "expectational paradigm" is the way in which the decisions of agents are influenced (1995:298). The more a social system enhances its complexity, the greater its ability to condition the expectations and behavior of actants in ways that do not appear constraining (1995:319). *Thus the interpenetration of social systems of domination, might result in an obscuring of the function and specificity of exclusion, oppression or exploitation because of the cross cutting difference within domination.*

The complex relations and formations of difference within structures Luhmann describes as a "matrix." I think that the notion of a matrix is useful to capture the complexity of the interpenetration of social systems of domination. "Hypercomplexity" describes a system that deploys differences in internal structuring and external interpenetration. This may capture the interlocking sense of dominations of difference I am looking for, but in Luhmann's hypercomplexity, there is no way out. Morphogenetic change creates new structures and systems out of old; complex systems require conflict, instability and contradictions for these preserve the autopoesis of the system (1995:352–90). Luhmann is aware that his theory "does not recommend itself as a nice, cooperative one (but one which) is interested in the normalization of the improbable" (1995:394). A question Anne Oakley (2002) asks of the contemporary "state" of gender and people/planet relations is "how" we got into such a mess. The ability of social systems to embed us, to obscure domination through intricate complexity of inclusion/exclusion, multiplicies of difference and interpenetrative systemic logic is an answer, and applied more concretely to "real" social systems, we can tease out a picture of the web of domination. This a point ecofeminists have made for over 30 years – the interces of domination is strength. I am more of an "enlightened woman" (Assiter 1994) than perhaps I like to think. There probably is a way out, and although excavating such is beyond the parameters of this book, I consider an analytics that engages with a multiplicity of possible difference-in-dominations, is best placed to show the weakest links and points of possible unraveling.

Luhmann's own model is unashamedly anthropocentric. His understanding of social systems as those of communication of meaning, a phenomenon which does not exists "outside of society" (1995:34–7), means that "machines," "organisms," "social systems," and "psychic systems" belong to different strata, different arenas of life (1995:410–17) and there is the implicit assumption that they do not interact or "interpenetrate." What differentiates social systems from "organisms" is the deployment of meaning, and this renders "society" exclusively human (1995:46). This irreconcilable difference of species renders interpenetration of social systems and ecosystems impossible (1995:102). Luhmann is too keen to make conceptual distinctions between social and other types of system. The notion of interpenetration is useful for understanding the ways in which ecosystems are both systemic and bounded entities, whilst at the same time, implicated and embedded within human social systems. The environment is both characterized by a complex interlocking mass of multivariate and nested systems, and at the same time, ecosystems are interpenetrated by human social systems, specifically in the ways that nature is socially mediated and constituted under anthropoarchal relations.

The conceptual apparatus of complexity theory enables theorization of multiple and cross cutting difference within social systems resulting from interpenetration of various other social systems that form their environment. Luhmann himself says nothing of social exclusion and inequities, but he provides a framework for theorizing multiple systems of social domination, which interpenetrate, co-exist and co-evolve. A social system of gendered domination, patriarchy must be considered as existent in the context or milieu of a variety of other systems of domination based on difference. Patriarchy, like other systems of domination, is internally differentiated, and following Walby, I see this differentiation in terms of structures. What follows is a sketch of some structures of which a patriarchal system might be composed, in the specific context of the wealthy parts of Europe and North America, following Walby's suggestions. An elaboration of structures is a way of outlining boundaries, and helps us see less abstractly, the nature of both particular systems and their "points of contact" with other systems. The meshing of system boundaries might both increase internal complexity, and contribute to the reinforcement of systems of domination. Luhmann suggests that the increase in complexity in the elements composing a social system occurs when "it is no longer possible at any moment to connect every element with every other element" (1995:24). I would like to draw on this idea of incomplete connective complexity to argue that key to the understanding of the interpenetration of social systems, is the divergence in the structural forms which might be apparent in the "real" world. As will be apparent from the outline below, structures of patriarchy are both different to and overlapping with, those outlined in the previous chapter, concerned with a social system of domination over "nature." Following Luhmann (1995:27), I suggest that we shift to a "completely different understanding" of social systems "that must be formulated entirely as difference in complexity."

Structures of patriarchy

Patriarchy has been defined as a system of social relations based on gender domination in which primarily women, and also children and insufficiently patriarchal men, are dominated and oppressed, exploited and marginalized. Structures of patriarchy are systemically linked and are specific patterns or forms, emergent from normative praxis. Walby's 1990 model involves sets of relations embedded in institutions and practices across six sites: employment, household, culture, violence, sexuality and the state. Latterly, opting for the less controversial "gender regime," rather than patriarchy, she has also replaced the term structure for the concept of "domain" within which sets of social practices cohere. Domains are grouped into the economic (involving the market and the household), political (states and supra national organizations) and social ("civil society" involving sexuality, interpersonal violence and social movements) (Walby 2004:10). The specification of these forms is much the same, and I adopt the "original six" structures, whilst modifying aspects of their content.

Sexuality

Sexuality is constituted through the deployment of discourses based on power relations of gender, involving the normative praxis of presumptive heterosexuality and more generally, the sexualization of gender domination, particularly in popular culture. I would include fertility and reproduction as part of the structure of sexuality, the control of women's fertility being in part related to sexual behavior. There are links between sexuality and other structures of patriarchy. Outside the household, sexuality is prescribed and enforced via the agencies of the state, which may in some cases be restrictive of women's autonomy, or, as in the case of the European Union and some of its member states, may be tolerant of diversity. Sexuality is embedded in popular culture and with the decline of privatized control over women and female sexuality (presumptive marriage and monogamy), public modes of cultural control over women become increasingly significant. Not all groups of women are subjected to the same mode of control in the same way. Some Western women are still strongly subject to privatized controls in terms of labor, reproduction, sexuality and culture. Although in public patriarchy the sexualization of popular culture is an increasingly prevalent mechanism of control, this does not imply that there has been little change. For example, in Britain, whilst presumptive heterosexuality and active heterosex is more dominant in popular culture than 30 years ago, this operates alongside an increasingly diverse range of sexual representation.

The household

Despite radical changes in household norm and form, the continuing gendered division of domestic labor continues in heterosexual households, as does domestic violence against women. Black feminists have particularly critiqued conceptions of the household as "oppressive," because they do not sufficiently account for the difference of family form and function for black women engaging with a public sphere that is racist. Yet the household remains a privatized site of exploitative relations in terms of domestic labor and childcare, which are strongly gendered inspite or despite the differences of race and "culture," albeit the specific qualities of domestic work may take differing forms and be experienced differently by divergent constituencies of women. For example, in households where family members are able to provide childcare support, the responsibility for the care of small children may weigh less heavily. This experience is also divergent depending upon a woman's position in relation to paid employment. Those for whom the labor market offers restricted and perhaps undesirable options may have vested interests in homemaking as a full or part time occupation.

Paid employment

Women in various forms of patriarchal system are paid less than men and horizontally and vertically segregated in low status gendered employment. Whilst the relative exclusion of women from paid labor in "Western/Northern" areas of the globe is radically diminished over the last two centuries, gender segregation has declined only slightly. Patriarchal relations of production generally constitute women as a "cheaper" labor force than men, and they may be increasingly drawn into the labor force as a consequence of the restructurings of postindustrial capitalism and the demand for cheaper labor, yet they are still drawn in, in gender segregated ways. This pattern is also strong in poorer regions of the globe drawn into the division of labor of globalized corporate capital (Mies et al. 1999). Women remain more concentrated in low paid occupations than men, and material factors coalesce to structure women into such jobs. Walby sees the labor market as shaped by racist processes which impact particularly negatively on men from minority ethnic groups, although these forms of restructuring also have gender implications for workers in the poorest parts of the globe, where women are also constituted as cheaper and more exploitable labor forces with very limited political muscle. In Western late modernity, women's labor is less available for domestic exploitation, and exclusionary practices are increasingly limited by state intervention. Yet there are significant regional variations both at the local level of the nation state and an increasingly marked international division of gendered labor (Oakley 2002:98). In the early decades of the twenty-first century, we have seen an overall increase in women's participation in paid employment, combined in Europe for example, with the increased political regulation of the labor market with respect to gender inequality (Walby 1999).

Culture

Patriarchal culture involves the creation and deployment of complex sets of gendered discourses, and the representation of gender through specific institutions and processes. The latter refers primarily to the media and forms of popular culture (such as film, literature, advertising), and also to educational and religious institutions and the formal processes of education, and to institutions and processes of leisure. Contemporary Western discourses of femininity do not focus only on domesticity, but also on (hetero)sexuality. Discourses of femininity articulated in popular culture such as "women's magazines" have incorporated women's paid employment although they still often presume gender differentiation. "Men's magazines" in countries such as Britain can be seen as a retrenchment of traditional patriarchalism in popular culture, which ranges from antifeminism to outright misogyny (Whelehan 2000). Sexualized discourses of femininity constitute a female imperative of heterosexual attractivity and availability, whilst those of gendered domesticity suggest the normative desire for motherhood, and "care" for male partners and children. As more women contest and reject this domestic role, the cultural control of women has shifted toward gendered sexualization in popular culture. Discourses of masculinity are less altered. Whilst masculinity is no longer represented in and articulated through, the keeping of a dependent wife, paid employment retains its signification for heterosexual masculinity. Discourses evident in popular culture are important in the construction and reproduction of gendered power relations, and also involve important shifts in the reproduction of meaning.

Violence

Patriarchal violence takes various forms, from rape, child sexual abuse, the battering of female partners to less physically harmful instances of sexual harassment. Some feminists have rightly noted certain forms of violence are strongly gendered, although they do not involve a female human victim, such as: warfare (Pierson 1988, Mc Allister 1982, Enloe 1983, 1999), the environment (Warren 1994, Plumwood 1993), racism (Griffin 1981, Spiegal 1988) and animals (Adams 1990, 2003). My understanding of patriarchal violence extends to all groups that suffer systematic gendered and sexualized violences, and may take non-physical form, such as the threat or the fear of violence. In addition, violence can be suggested, for example in images in pornography, and other forms of popular culture. Discourses of masculinity include machismo, according to which it is appropriate for men (and some women) to use violence against each other and against "Others." Whilst some nation states and/or local states within them may have taken concerted action to limit the most extreme effects of domestic violence, this remains marginal as a concern at the national (Kelly 1999) or international (Hanmer 2000) level.

The state

The (nation) state retains its saliency as the formation of contemporary politics, and the limiting impacts of "globalization" are overstated (Hirst and Thompson 1997). The state is shaped by structural considerations pertaining to various systems of oppression based on gender, class, race, and nature, and the specific policy and institutional formations these interpenetrations give rise to are timealized and spatialized. The state may articulate patriarchal domination via its non-decisions and apparent neutrality, as radical feminist have contended for example in non intervention in cases of domestic violence (Kelly 1999), and structural bias in law (Mackinnon 1989) particularly in cases of femicide (Lees 1990, Radford 1994) and rape (Kelly 1988). In the last quarter of the twentieth century, largely in response to feminist political action, the state in "liberal democracies" shifted policy regarding gender relations, resulting in some benefits for women, and various strands of feminism have been seen to have some impact on policy making. However, certain other policies have, in an indirect way, had negative effects on women (such as cuts in welfare provision). The role of the state in reinforcing gender relations can be seen largely in its lack of intervention to protect women and act against inequalities, which legitimate the patriarchal status quo, particularly around childcare, and configurations of political parties and policy makers are biased against radical restructurings of gender relations, "state feminism" notwithstanding (Oakley 2002:45). Walby's own position on the state has shifted significantly. Whilst she argues that the nation states are increasingly seen as mythic yet powerful entities (Walby 2003c), the increased political representation of women, particularly in the context of the European Union has led to a series of positive initiatives in the interests of many women as workers and parents (Walby 2003b:7).

Walby's model of patriarchy has been influential yet controversial in feminism strongly influenced by the "poststructuralist generation" (Braidotti 2002). The most common specific criticism is a version of Anna Pollert's (1996) assertion that gender relations are "everywhere," and their analysis within six structures is arbitrary "Why not four, or forty, or whatever?" (1996:645). Walby's selection is based upon certain "sets" of gendered relations articulated through closely interrelated groups of institutions and processes that capture relatively discreet arenas of women's experiences and can be seen as key sites within which certain oppressive relations cohere. Walby's shifting conceptualization may be taken for some as an indication of the arbitrary nature of such classification, yet the reconceptualization constitutes a refinement rather than a substantive change and an attempt to deepen the complexity of structural analytics. In many ways, Walby exemplifies the kind of approach to systemic relations that Luhmann suggests. Patriarchy is a network of relationships, in which connections are complex, and certain elements are connected to others within the system whilst other parts are not always linked. In her later work, Walby is particularly keen to pull through the fracturings of difference (or for Luhmann, differentiation) within her systemic structures. She argues that we can see complex patterns of differentiation where gender is cross cut by relations of age, ethnicity and class and where patterns of gender relations also adopt regionally specific formations. Like Luhmann however, her analysis does not attempt to account for the embedding of social relations in ecosystems, nor for the social relations of "nature." Both these sociologists operationalize a traditional conception of a bounded social system that does not interpenetrate non-social systems.

First world feminism has long struggled with the implications of the cross cutting social dominations of class and gender, and critiques of the eurocentrism of such theorizing has encouraged feminist theory to get to grips with multiplicity and complex fracturing of domination. It is still relatively rare for feminists to acknowledge their anthropocentrism. The influence of postmodernism has led to the "disrupting" and fragmenting of many categories of social and political theory, but the category human has been resilient to such deconstructive zeal. There are some who have recently made links between patriarchy, capitalism and environmental problems. Oakley (2002:123–44) for example, considers global warming and genetic modification and transnational corporate biodiversity prospecting, and its environmental and social impact, particularly on communities in poor regions of the globe, and particularly on women. Ecofeminism has sought to foreground such inter-relations, but in so doing has lost some of the *social* complexity grasped by black and socialist feminisms. I want now to pull through the understandings of patriarchy, capitalism and racism as systemic formations of social domination, with respect to the corpus of ecological feminist theorizing in seeking to further elaborate the case for multiple systems analyses. Keeping hold of the analytics of multi-faceted difference-in-domination, Chapter 5 takes ecofeminism on a complexity spin.

I have argued for a structural approach to the analysis of gender relations that operationalizes a conception of a system of "patriarchy," whilst recognizing that the interpenetration of multiple systems of domination means that specific forms, practices and collective experiences will demonstrate multifaceted complexities. Critiques of patriarchal theory are pertinent in indicating a lack of specificity in particular analyses, but they are themselves lacking in specificity and complexity if they suggest that these traits are inevitable or endemic in the adoption of a systemic and structural approach to gender relations. Post structural analyses are problematic in that their conceptual fragmentation and emphasis on agency fails to capture the systematic nature of gendered power relations. I have drawn upon critical realism in arguing that these approaches are reductionist in their almost exclusive focus on human agency in social relations. Gender relations may most usefully be seen to articulate in institutions, processes and procedures that can be conceptualized as structures. Such structures have certain effects and can be considered real, and may be both trans and counter phenomenal. Yet people have choices and options and may act as agents of reproduction within patriarchal structures or may also contest and change them. Agents are implicit in normative praxis, and restructure social institutions, practices and relationships, and thus are systems autopoetic and dynamic. Patriarchal relations and formations are historicized and spatialized. These are not virtual structures (Lopez and Scott 2000:94) but embodied in bodies, concretized in institutions, played out and played with in relations of everyday life. It is the intersection of relational systems that solidifies and strengthens structures of domination. Out of the apparent chaos of difference, comes the complexity of dominatory power formations.

5
Ecofeminism and the Question of Difference

> I often hear people arguing about the world's many evils and which should be the first confronted. This fragmentary approach is itself part of the problem, reflecting the linear, hierarchical nature of patriarchal thinking that fails to grasp the complexity of living systems. What is needed is a perspective that integrates the many problems we face and approaches them holistically.
>
> Petra Kelly 1994 (1997:119)

Ecofeminists draw upon deep ecological theory to the extent that they conceptualize human relations with "nature" as a form of domination. They also provide a version of social ecology in which the domination of nature is interrelated to intra-human social hierarchy and difference based on gender, race and class amongst other formations. Ecofeminism can be seen as a paradigm for the tracing of interrelations between different formations of domination based on difference. This chapter outlines key strands and themes within ecofeminist thought, and argues for a different framework for ecofeminist theory based on an understanding of complex systems. I suggest that ecofeminism needs to engage more fully with debates on the efficacy of patriarchy, and to be rather more generous toward deep and some socialist ecologies with respect to their analyses of capitalism and anthropocentrism as systemic relations of domination.

From the early years of the "second wave," some feminists were suggesting that environmental degradation was a matter for feminist theory. Feminist involvement in anti-militarist politics and a plethora of environmental issues indicated that in praxis, activists were already engaging with these matters in feminist ways. As a body of social theory, ecofeminism has followed Simone de Beauvoir's suggestion that women have a particular affinity with the natural world due to their common exploitation by men (Mellor 1992:51). Unlike de Beauvoir however, all the differing ecofeminisms have argued for the positive revaluation of the connection between women and nature (Plumwood 1993:8–9).

Victoria Davion (1994:8) suggests there are two strands of ecofeminist theory, but that one strand is distinguished by being "eco-feminine" and thereby not feminist. Chris Cuomo distinguishes "eco-feminism" which is primarily concerned with the similarities among the "objects" (such as women, animals) of oppressive thought and action (1998:6), and "ecological feminism," which focuses on the links between forms and instances of oppression (1998:7). Cuomo overstates this difference and in the work of individual theorists both are usually attended to. Mary Mellor describes the difference in terms of "social" and "affinity" explanations of women's relationship to nature (1992:51–2), and these are the terms I tend to use. I do consider this division rather arbitrary however. Affinity and social ecofeminisms have many common concerns, including violence against women and often against animals, women's health, the gendered division of labor in production systems and the household. They have questioned forms of social inequity resulting from the exploitative relations between rich and poor countries, and see human domination of the environment as related to a worldview that justifies the domination of women.

The difference between social and affinity ecofeminists is the latter's emphasis on spirituality, and the physical bodily experiences of women, which encourage identification with "nature." Social ecofeminists tend to focus on ecofeminist ethics and engage more closely with green social and political theory. Affinity ecofeminists are most closely related to radical feminism and tend to see patriarchy as responsible for our currently destructive relationship to the earth (Leyland 1983:72). They have developed theoretical links between feminism and ecology with specific reference to sexuality, motherhood and reproduction, warfare, and male violence. This writing has been criticized as essentialist, for apparent allusion to the particular knowledge, emotion, sensuality, thought and morality of women (Segal 1987). Within ecofeminism itself, some have accepted the "straw woman" as a means of distinguishing their own work as non-essentialist. There is merit however, in the range of ecofeminist approaches and the allegations of essentialism, are, more often than not, based on cursory readings, de-contextualization and attribution. Critiques of ecofeminism are often startling examples of academic foul play. This said, most of the approaches discussed are criticized for homogenizing dominations based on gender and nature as the product of one all embracing system of oppression, patriarchy, or, less usually, capitalism. It is argued such homogenization prevents us from capturing the complexity of oppressive relations that might be captured by a multiple systems approach.

There are also difficulties with the ways some ecofeminists have deployed a concept of patriarchy. In some cases, the use of the notion of patriarchy has been underdeveloped and used in a rather careless manner. For example, in introducing an early anthology, Judith Plant uses the term loosely to apply to both some undefined object, and to the gendered division of labor:

> neither the present understanding of what is female or male is an adequate characterization of what it is to be human. For both genders are fraught with pathological behaviours which serve to perpetuate the system of domination ... The message of ecofeminism is that we must all cultivate the human characteristics of gentleness and caring, giving up patriarchy with all its deadly privileges ... This means valuing diversity above all else. (Plant 1989:2–3)

It is not clear from this account what a patriarchal society *is*, or how it systemically privileges some groups of humans. Most ecofeminists tend to associate patriarchy with the division between public and private spheres of life, in which the private sphere is associated with nurturance and the reproduction of everyday life and is both devalued and associated with women. In addition, this is environmentally problematic, as men are further away than most women from the reproduction of the material conditions of life. However, ecofeminism needs to elaborate and defend an understanding of patriarchy, and make explicit in what ways it overlaps with, and is distinct from, other forms of social domination.

This chapter intends to draw out latent elements of systemic theory in many ecofeminist accounts. The first section considers women's difference from men in ecofeminism, concentrating on what I see as discursive analyses of culture, investigating symbolic regimes, value systems and religious myth. These perspectives illustrate some of the difficulties with the "greening" of what is implicitly a theory of patriarchy, but I also dispute the particularly strong critiques of these ecofeminisms as essentialist. Second, I examine more historically grounded ecofeminist accounts that include both symbolic and material change, but do not explicitly take on board the structural and system implications of their analysis. Finally, the work of those who explicitly theorize the patterns of connection between the domination of nature and different forms of social domination are considered, and I make my own case for a complexity approach to multiple systems.

Women's affinity, men's distance? burning the straw women

It is not only those influenced by theories of postmodernity in the social sciences who have "turned" to cultural analysis, but a whole range of thinkers with differing epistemologies and ontologies. "Affinity" ecofeminism is often associated with radical feminism and Western neo-paganism. In part, this connection stems from the dominance of American radical feminist thought in the emerging ecofeminist positions of the 1970s and early 1980s. This has meant ecofeminism has been perceived to be a reflection of the ideas of some particularly controversial figures, such as Mary Daly and Susan Griffin, who have been described by critics as "cultural feminists" (Evans 1995). Yet many ecofeminists implicitly deploy some form of discursive approach to the investigation of relations between gender and nature, and this merits some (re)consideration.

Discourses are interrelated sets of ideas, which concretize themselves in specific practices, processes and institutional formations. They are part of the architecture of relations of domination, or for Foucault, technologies of power. I undertand them as reflective of the detailed and variant formations of power evidenced in multiplicities of difference and as constructing and reconstructing systemic patterns of domination. Foucault analyzes power as operating through the "functioning of a discourse," the multiplicity of discourses constructing "relations of domination" (1976b:31). At various times there have been important shifts in "discursive fields of knowledge," and changes in predominant discourses indicate how the content of discourses structures the ways in which individuals think and talk about objects of investigation. Discourses constitute forms of power and domination, they discipline the subject in unseen ways, and are not merely sets of ideas, but also institutionally rooted social practices which structure the social world and which have "real effects" (1976b:35). Although discourses are dynamic and heterogeneous, they are produced within an "episteme," a condition of possibility (Foucault 1972). I think there is a case to be made for Foucault capturing a notion of structure here, within which discourses operate. In Foucault's earlier (and later) work, discourses are conceptualized as structuring forces, having power to constitute (but not determine) both ideas and their related social institutions and practices.

For example, in his earlier study of madness, Foucault's use of discourse is linked to systemic domination, in particular capitalism. In the context of the emergent power relations of capitalism, religious and Enlightenment philosophical and medical ideologies transform a socially tolerant conception of madness, to one in which insanity becomes equated with the "uncivilized." The shifting symbolization of madness had direct and concrete effect on the treatment of those identified as "mad" who were confined and brutalized within asylums and workhouses (1971:61). This conception of oppressive discourses of power which structure social life is also evident in his work on punishment and the development of prisons, wherein deviation from capitalist values was punishable by a particular form of discipline and surveillance (Foucault 1979). It is this discursive realism that enables a foucauldian reading of some ecofeminist writings. The identification of common discourses may suggest specific forms and degrees of power relations between/across gendered and natured domination, and this in my view, is how some ecofeminists have used discourse as an analytic tool.

Patriarchal discourses of gender and nature

A number of ecofeminists have suggested that patriarchal discourses carry gender dichotomous normalizations, which feminize the environment and animalize women, constructing a dichotomy between women and nature, and male dominated human culture. The arguments presented often also draw on a form of standpoint epistemology: gender roles constituted through such discourses render women in closer material proximity and relation to the environment than men, and some suggest further this means a greater potential to develop an ecologically sensitive value system. Some thinkers also contend that a new culture based on re-valuation and radicalization of certain "feminine" qualities, can contest the ecologically destructive system of patriarchy.

Patriarchy deploys different discourses of masculinity and femininity that associate women with nature and men with culture. Patriarchal culture venerates masculinized qualities such as virility, strength, self-control, emotional reserve, competence, rationality and devalues feminized qualities of caring, sensitivity, fragility, dependence, emotionality, tenderness and sensuality (Lowe and Hubbard 1983:2–3). Some of these gendered values are seen to be patriarchally contesting. The normalizing processes of patriarchal society, some argue, leads to a gendering of the respect for life (Freer 1983, Elshtain 1987, Ruddick 1990). By contrast, according to Griffin (1984, 1994) and Daly (1979, 1988), patriarchal culture venerates death and violence, and is preoccupied with dominance and control over women and nature. Ynestra King contends that ecofeminism is about the connectedness and integrity of living things (1983:10–11). Patriarchy enshrines a hatred of women and nature, and this "masculinist mentality" is responsible for environmental devastation. In many cases, this worldview is predicated on a denial of human embodiment. As King suggests:

> All human beings are natural beings. That may seem an obvious fact, yet we live in a culture that is founded on the repudiation and domination of nature. This has special significance for women because, *in patriarchal thought, women are believed to be closer to nature than men*. (1989:18, my emphasis)

Whereas King suggests the connections between women and nature are symbolic constructions, others consider that women may empathize with others due to their social role, for example, as "mothers" and potential mothers (Freer 1983). This is the basis on which some theorists posit a gendered concern with the treatment of animals. Connie Salamone (1982) claims that women's social practices of care for others means they are more likely than men to oppose practices of harm against non-human animals. Josephine Donovan and Carol Adams (1996) have proposed a new basis for animal rights theory by arguing that women have a sense of responsibility toward animals deriving from their praxis of "caring." Norma Benny contends that many women can empathize with the sufferings of animals, as they have some common experiences, for example, female domestic animals are most likely to be "oppressed" via control of their sexuality and reproductive powers, involving varying degrees of physical violence and emotional deprivation (1983:142). Carol Adams (1990) argues meat eating is a masculinized and natured practice predicated on the gendered and natured oppression of "meat" animals, and that popular culture is saturated with interpolations of gendered nature, and natured gender (Adams 2003). Thus some consider that gendered and natured normalization captures animals and women, in some instances, within the same discursive regime, and may place women in a position of possible contestation regarding the treatment of animals and the eating of meat.

One of the best-known ecofeminist writers is the poet, Susan Griffin, whose *Woman and Nature* (1984) traces a history of patriarchal thought concerning nature and women. Griffin juxtaposes two sets of discourses throughout the book. One set contains dominant Western narratives on women and nature, exemplified by the teachings of Judeo-Christianity, and the understandings of mechanistic science. Common to patriarchal narratives is a conceptualization of the natural world as transient matter, and an association of women as closer to nature due to reproductive capacity and sensuality. In this discursive regime, women become part of nature and distinct from culture (see also Figes 1970). Patriarchal discourse, whether sacred or secular is gendered – its discourses normalize rationality for men, and intuition for women. Griffin considers practices of human domination and control over the environment, arguing such control has been conceptualized in terms of gendered sexual possession. She implies for example, that elements of the process of domestication of animals and women have a similar history – premised on feminization, sexualization and control of sexuality (1984:66). Her method of presenting the historical and contemporary experiences of "nature" and women is as a dialogue of voices designed to indicate the slippage of terminology applied to women and both "wild" and agricultural animals, ocean life, forests and so on. The possibility of slippage, the normative application of nature to women and female to nature, illustrates the duality of the discursive co-construction of "women and nature."

The nemesis of postmodern feminism – Mary Daly, considers that "Patriarchal society revolves around myths" (1979:37) of male dominance which are oppressive for both women and "nature." These myths are sets of ideas that carry relations of power and have a real effect in structuring social institutions and practices. As such, they can be read as discourses. These "myths" are reflected and reproduced in social practices and institutions, and dominant relations of power often compel people to act within such discourse (1979:46). The concretization of patriarchal discourses in institutions and practices is a "sado-ritual" syndrome involving the infliction of violence against women, which Daly illustrates with five historical cases, drawn from different centuries and cultural locations. Unfortunately, she concludes the same patriarchal processes are evident in all five rituals (1979:394), marginalizing elements of difference either in the content of discourse or in the form and degree of violence across time and place. Patriarchal sado-ritual is an explanatory model for a range of violent practices, including environmental damage, and Daly suggests that women and "nature" inhabit the same space, the "Background" (1984, 1988), the arena of reproduction of species and embodied existence, in contradistinction to the "foreground" (1979:26) of public patriarchy. Background experience enables women to empathize with nature and "Trees, Stars, Animals of all Kinds" are "companions" on "the Wicked Weaving Journey" into post-patriarchy (1988:90). Patriarchy and environmental abuse are challenged, for feminist thinking is ecological ("gyn/ecological," 1979:21). In seeing patriarchy as paradigmatic for other kinds of domination however, Daly ignores qualitative differences in relations of domination that are produced by the intersection of gender with "race," class and nature and other formations of difference-in-domination.

Whilst neither Daly or Griffin utilize the term "discourse," they have such a conceptualization – sets of ideas that are infused with relations of gendered power, and have real effect. This is also clear in Griffin's work on pornography (1981) and war (1994). Daly is more explicit in her identification of particular structures of patriarchy (1979, 1988): reproduction, employment, political institutions, the media, warfare and religion. Neither Daly nor Griffin can be labeled biologically essentialist, although they may underestimate the cross cutting influences of oppressions based on race and class as qualitatively and quantitatively affecting the form and degree of women's oppression. This tendency to homogenize is resultant from an all-encompassing systemic understanding of patriarchy as a system of "Othering." Adams and Donovan (1995:3) have similarly contended that patriarchy is "prototypical for many other forms of abuse," and Suzanne Kappeler (1995:348, also Collard 1988) has echoed Daly and Griffin in asserting that patriarchy is "the pivot of all speciesism, racism, ethnicism, and nationalism." Yet patriarchy alone cannot explain other forms of oppression, exploitation and domination or account for differences in forms and degrees of domination. In addition, there are difficulties inherent in a "reversal strategy" of social change in which the dominant culture is subverted by giving positive value to what has been previously despised (see Plumwood 1993:30–3). The dualisms of humanity/nature and men/women are so closely intertwined that we must theorize a way of transcending them both, rather than reinforcing the initial dichotomy. Despite such difficulties, these approaches provide a powerful analysis of the ways that social exclusion, particularly the gendered division of labor, structures relations between society and the environment, and of the ways narratives of the domination of nature are interwoven with those that marginalize certain social groups, and, in turn, are embedded in institutions and practices.

Enchanted ecofeminism

John Barry (1999:17) has argued that deep ecologists seek a "reinchantment" with nature, given that the processes of modernity have resulted in its disenchantment. In this process, as we saw in Chapter 2, some have drawn upon pre-industrial societies and dug into pre-history in order to find evidence of a more appropriate relationship to nature. Others have dabbled with neopaganism and non-Western religions in order to foster their ecological selfhood. In similar vein, some ecofeminists emphasize the political significance of women's earth-orientated spirituality in challenging the social norms and values of patriarchal religion. Such "enchanted" feminism and ecologism (Salomonsen 2002) has been strongly emphasized by critics, who claim that these writings are highly "essentialist" in implicitly presuming an authentic self. I consider such claims to be overstated. They ignore divergences in perspectives and the possibility that even an "essential self" may also be able to account for difference.

Rianne Eisler (1990:33) sees ecofeminism as reaffirming an earth-orientated religion of pagan goddess-worship. Carol Christ (1992:277) argues that this legitimates female "power and authority," and for Wiccan practitioners such as "Starhawk" (1982), paganism celebrates women's bodies, sexuality and our embodied condition as human animals embedded in "nature." Vandana Shiva (1989) has contended that aspects of Hinduism celebrate natural diversity and female spiritual strength, and are inspiring images for ecofeminists, although she has been criticized (Agarwal 1992) for failing to point out that other aspects endorse female subordination and the strict social hierarchy of caste. Critics have argued that merely adopting more positive myths and symbols for women will not change the reality of social institutions and practices (Biehl 1991). Charlene Spretnak (1990) has responded to such charges, plausibly perhaps, by positing that such mythologies are indirectly politicizing – they may inspire and thereby encourage political action to change social reality.

Ecofeminist forays into pre-historical anthropology have tended to posit a transition from gylany (Gimbutas 1982:17), gynocentric society (Starhawk 1990b), or matriarchy (Reed 1976), to patriarchy. The establishment of patriarchy was a takeover of spiritual, political and social power that was gendered and natured. For example, mainstream anthropology has generally seen Paleolithic society as characterized by male dominance through game hunting and has upheld the assumption of a patriarchal society (Fisher 1980:136–7). However, some feminist scholarship sees the late Paleolithic as likely to have been characterized by an animistic worldview, providing the foundation of the Goddess religion (Lerner 1986:148–50, Stone 1977:28–9). For Marija Gimbutas, archaeological evidence suggests Goddess worshipping Neolithic societies lived in harmony with nature, and that reverence for nature and women was linked (1990:321). Excavations have suggested civilizations in which women participated in religious life, were priestesses and other social leaders (Mellaart 1965:86–8), matriliny was the norm (Stone 1977:49–54), and sex was sacred (Lerner 1986:103). These theorists collectively argue that in Goddess worshipping animistic societies, women, animals and sexuality were seen as sacred, and this was reflected in an absence of gender stratification. What happened to such societies is even more speculative, but a similar thesis of a religious "fall" is often advanced, wherein the Near East suffered cultural disruption from invasion by migratory waves of northern Indo-European, Aryan and Semitic pastoralists (Gimbutas 1977:293; Lerner 1986:162–3), whose religion venerated hunting, war, sacrifice, and an omnipotent male god (Stone 1977:82). Their "dominator" model of social organization revolved around developing technologies of warfare (Eisler 1989:45) and hierarchies of kingship, class/caste, slavery, and gender (Starhawk 1990b:46). Patriarchy and human dominance over nature did not emerge with modernity argue these theorists, but with pre-historic change from Goddess worshipping animism, to male dominated hierarchical religions.

Others have drawn upon early modern connections between gender, nature and neopaganism looking at the European witch burnings of the fifteenth to seventeenth centuries. Carolyn Merchant (1980:140) argues European witchcraft was an historical "reality" and its belief system was conceptually animistic. The witch burnings themselves have been seen to exhibit a number of gendered and natured relations: sexualized violence ("gynocide," Dworkin 1974, Daly 1979), violence against animals (as "familiars") (Starhawk 1990a), criminalization of gynocentric herbalism (Ehrenreich and English 1973), the attempt to eliminate pagan animism (Eastlea 1981:111–42) and female social and sexual (Starhawk 1990a:207) independence. The focus of such work is to elucidate the symbolic and social position of women, and how this changes when the social and natural worlds are dichotomously conceptualized. Further, there is a belief in political efficacy. Starhawk for example, believes that profound social changes are closely linked to shifts in religious symbolism (1990a:72, Freer 1983:131), and this is a theory of symbols which is derived in part from a reading of Mary Daly's early work. These readings of pre-history are part of a mythopoetic contribution to theology and do need to be differentiated from the work of those who do not sit within the theism (like Daly 1979). In addition, spiritual ecofeminists go further in asserting the power of symbolic regimes to define reality.

Neo-Pagan ecofeminism is "essentialist" to the same degree as any other religious philosophy, in that it posits normative truths about social life. It has a notion of the "deep self" (Salomonsen 2002:182) with clear similarities to deep ecological conceptions of selfhood, for it posits the concept of "immanence" (spirit, which runs through all life, but is concentrated in organic matter, and can be channeled by sentient beings, Starhawk 1990a:136). There is a holistic conception of spiritual embodiment in nature, and of a "true self" corrupted by patriarchal and other oppressions (Greenwood 1996:111). The equation of ecofeminism with such spirituality has been challenged with the marginalization of women's difference, such as the positive and empowering Christian experiences of many black American ecofeminists (Taylor 1997). However, Salomonsen (2002:125) suggests that paganism and Christian/Jewish understandings of the world are not necessarily opposed, with feminists increasingly adopting a plural spiritual identity (see Radford-Ruether 1983). Paganism itself acknowledges multifarious forms of spiritual manifestation, multiple truths and sees divinity as inseparable from, and immanent in, nature (Crowley 1989), and its animism impels ecological consciousness, for if we see all things as connected, we have a political responsibility for the natural world (Adler 1986:410). Starhawk presents her enchanted ecofeminism as a political project based on a respect for a "multiplicity of spiritual directions" (1989b:183) and valuing human and non-human diversity.

Magical ecofeminism has made an important contribution to the praxis of ecofeminism, and certain influential writings have been taken out of context by feminist social scientists. These are elements of an emerging theology that have drawn from, and influenced, academic theorizations. Secular critics take the apparent essentialism in such writings too seriously. I would not want to define magical ecofeminism as apolitical or as theoretically irrelevant. Mary Mellor (1997:146–7) has argued that a "politics of nature" requires a "deep" view of humanity as embedded in nature and interconnected and interdependent with the natural world. She endorses Starhawk's (1990a) view that a radical politics demanding dramatic changes in socio-economic formations of exploitation and exclusion can coincide with, and be enhanced by, a spiritual awareness of human immanence. Such awareness does not imply a mystical belief in a transcendent power, but can be understood as a deep level of human ecological awareness. Mellor's suggestion is an "immanent realism" which conceptualizes the relationship between humanity and nature as emerging and shifting dialectically, and "realized materially as a living process" (1997:149). Like Mellor, I see ecofeminist neopaganism as politically non-problematic, whilst also being most interested in the material formations of domination and the contradictions of embeddedness and difference captured by more secular approaches.

Burning the straw women

When it comes to critiques of ecofeminism, Griffin and Daly have often been set up as the straw women – they, and one each of their many books, constitute ecofeminist theory. Stacy Alaimo (1994) misreads Griffin as glorifying patriarchal gender roles, whilst redefining self-ascribed ecofeminists such as Ariel Salleh and Merchant as critics of ecofeminism, on the dubious grounds that Alaimo actually agrees with some of the things they say. Carol Stabile considers Collard, alongside Griffin and Daly and suggests that they all draw uncritically on associations of femininity with "the primitive or the premodern," and "accept a stereotypical rendering of femininity" (Stabile 1994a:60). Yet Griffin and Daly are both highly critical of the impact of patriarchal femininity on women, and some of the images chosen for feminist identification are utterly antithetical to "stereotypical" femininity – man-eating lioness for Griffin for example, and axe hurling Amazons and Furies for Daly. Cuomo (1998:120) rightly suggests that Daly and Griffin articulate "unimaginable" ontological possibilities within patriarchal normativity.

Daly is particularly criticized for presenting a depressingly static view of gender relations, yet her work demonstrates an incredible optimism in the power of women to change themselves and patriarchal structures. Daly's "alternative culture" is an elitist notion from which men and most women are excluded (Segal 1987:21), but this is but one vision of post-patriarchy and Daly recognizes (1984:4) and positively regards (1991:xxxii), profound diversity amongst women. Her most recent offering, *Quintessence* (1999) can be read entirely as an exuberant discussion of women's empowerment and agency.

Griffin's apparent concentration on empowering qualities of motherhood has been much criticized for denying the variety of women's experience (Segal 1987) and sentimentalising domesticity (Soper 1994). The domestic sphere remains a site of exploitation of women's labor, and it is this which leads to the devaluation of women and the values associated with femininity (Dephy 1987). Griffin (1994:167) is clear however, that it is gendered discourses that normalize women's caring labor, and a positive reading of female domesticity is but one amongst many themes in her early work. In more recent writing, Griffin (1994) has questioned the notion of domesticity as female and privatized arguing that world changing historical public events, like war, are "embedded in us," constituted through and reflected in our supposedly private lives and spaces. Nearly all the critiques of Griffin refer to *Woman and Nature* first published in America in 1978, which is an intellectual work, but not an academic text. It is a piece of creative writing, or "art" (Rich 1982), and we must consider this specificity. My reading of Griffin's work taken as a whole is of a challenge to dichotomous thinking, and of mapping the connections between relations of domination based on assumptions of difference. Key to all this is the materialist notion of our embodied and embedded existence, and the differences and oppressions of gender, of "race" of nature, of the dichotomous spheres of public and private in Western social and political thought and practice, that are linked to a denial of embodiment (see Griffin 1989:7-17).

Whilst Griffin ignores the possibility men may connect to nature through their bodies, she does not exclude it. Her point is that socially constructed gender roles constrict women to a more embodied materiality. To contend, as does Susan Hekman (1990), that we must exclude biology from social explanation, is nature-phobic and *socially* essentialist. In arguing that women associated with a disproportional amount of caring and domestic labor, have a privileged understanding of the problems with our relationship to nature than those who are relatively advantaged by having less experience of nurturing work is an articulation of standpoint epistemology, and for some, this warrants the label of social essentialism (Ferguson 1993). Whilst Daly and Griffin do underplay differences between/among women, this is not an inevitable product of conceptualizing women's experiences. As Amber Katherine (1998) has pointed out, the often-cited critique by Audre Lourde in her "Open Letter to Mary Daly" does not charge Daly with essentialism (indeed, Lourde's own contribution to ecofeminism can be read as essentialist to the same degree, Cuomo 1998:122), but with universalism. These are not the same things, yet many critics conflate them.

Some have read Daly and Griffin to be endorsing "goddess" spirituality, a reading which manages to ignore the irony and humor within their work, and particularly in Daly's case, a play on religious iconography (see 1991:418-22). There is particular hostility to spiritual ecofeminism, as reinforcing cultural stereotypes, and Haraway asserts she would "rather be a cyborg than a goddess" (1991:181). Whilst I have concurred that ecofeminist Wicca involves an "essentialist" conception of the self, I am not convinced this necessarily involves the eclipse of difference – that self is arguably a post-patriarchal contestationary self, which celebrates diversity. Any essentialism is more likely to be a result of an epistemology of religious belief, rather than in any homogenizing ontology of gender or other relations of social difference.

Perhaps more than any other variety of feminist theory, ecofeminism has been subjected to a sustained and harsh critique. Teresa de Lauretis (1990) has forcefully argued that the form and function of anti-essentialism within feminism has been a political smokescreen, and it is time to "up the anti" as she suggests and question the simplistic certainties of "anti-essentialism." Ecofeminism is engaged in providing a set of contesting discourses to describe and rescribe our gendered relations to nature. I use the term rescribing because I think they attempt to do just that, and that such rescribing is not ontologically "essentialist." The questioning and problematizing of our prevailing notions of gender, of the links between these and what we think of as nature opens up conceptual possibilities for new definitions and concepts.

Ecological embeddedness and the structuring of social difference

Some theorists have sought to map historically, the ideological links between gendered and natured domination and the social institutions and practices though which such marginalization, exploitation and oppression takes place. Such approaches have also often been discursive, examining the shifting and interwoven narratives on women and nature that accompanied the transitions to modernity in Europe, and their concretization in specific institutions and material practices.

Capitalism, colonialism and the gendering and naturing of modernity

Carolyn Merchant (1980) identifies mechanistic science as a key structure in the control, domination, and exploitation of women and nature, and she attributes the "death of nature," and the move from a gylanic to patriarchal society, to the scientific revolution. Vandana Shiva (1988) focuses on Western modernity in terms of the impact of development as a process characterized by gendered and natured relations of domination. As argued with respect to Daly and Griffin, Merchant and Shiva do not use the term discourse, but their accounts capture a sense of discourse in their investigation of the concrete impact of shifts in paradigm that accompany modernization.

Merchant (1983) provides a critique of the scientific revolution of seventeenth-century Europe. She argues that both women and the natural environment were objectified and characterized as possessing similar subordinate inherent qualities, as a prerequisite for the commercial exploitation of natural resources and the social exclusion of women. Mechanistic science is a patriarchal discourse, and its development as a new intellectual paradigm led to the debasement of women and nature. Modern Western philosophy, she contends, has constructed a dichotomy between nature and male dominated civilization. The latter, in the guise of rationalism, scientific and technological development, has been responsible for defining women as the second sex, establishing a hierarchy of species involving oppressive relations, and legitimating the human domination of the natural environment. Merchant suggests that mechanistic science sanctioned the exploitation of nature, unrestrained commercial expansion, and a new socio-economic order subordinating women. She traces the decline of an older, animistic and gynocentric European worldview based on co-operation between humans and nature, and claims we need to rediscover such pre-modern ideas as a solution to the present environmental crisis and as a means of patriarchal contestation. The relationship between social exclusion and commercial exploitation is later elaborated (Merchant 1985) to focus more closely on the development of capitalism and its relation to the sexual division of labor. There are clear links between Merchant's ideas and those of eco-socialists. Men are associated with commercial production and women with unpaid labor and reproduction, and commercial production is also "alienated" from the natural world, which it pollutes and exploits.

Shiva (1988) examines the impact of Western modernity on "underdeveloped" countries arguing that the oppression of women and nature is linked, and discourses of modernity are patriarchal. Shiva's solution to the oppression of women and nature is similar to Merchant's; we must rediscover pre-modern, non-Western conceptions of nature and gender. Women in the Indian context already have such a conception, the holistic "feminine principle" that emerges from their daily practices as partners with the environment in food production. Shiva links the historical development of social, political and economic structures of colonialism to those of patriarchy and environmental exploitation. The

West has imposed its model of modernity on the rest of the globe through an ideology of scientific knowledge and the material institutions and practices of industrial capitalism. This has been ecologically destructive (particularly in terms of unsustainable agricultural practices), and has a gendered impact, excluding women in developing countries from their traditional roles in food production and subjecting them to invasive and inappropriate reproductive technologies. Thus for Shiva, the domination of colonized peoples, of humans over the environment and of men over women is linked.

Merchant and Shiva prioritize patriarchal relations in accounting for human domination of the environment. Whilst Merchant does see capitalism as significant, and Shiva emphasizes relations between patriarchy and postcolonialism, they do not unpack sets of oppressive relations but homogenize them. Shiva (1989:82) goes as far as to assert that "gender subordination and patriarchy are the oldest of oppressions," and argues that these structures of power operate to more severely oppressive degrees through the process of development or rather, "maldevelopment." An important strength of Shiva and Merchant's work however, is an implicit notion of systemic and structural domination, which is more developed than that found in Griffin and Daly. Shiva and Merchant use a form of discursive analysis to exemplify the content and patterning of such domination. They are concerned with alterations in predominant sets of ideas about gender and nature that are infused with power relations and have a real effect on social, economic and political processes, practices and institutions.

The gendering and naturing of scientific knowledge

A key thread running through the work of Shiva and Merchant is the role of Western scientific knowledge, institutions and practices as key to the structuring of gendered, racialized, natured and postcolonial social formations of domination. All but the most postmodern ecofeminists tend to have a realist epistemology, whilst at the same time, being critical of the gendered and human dominant content of much scientific knowledge. Ecofeminists have drawn upon feminist critique in seeing the development of "mechanistic" science as both gendered and natured.

For Shiva (1993) women's traditional ways of knowing are undermined by the Western rationalist paradigm, establishing a "monoculture of the mind," which is both an aspect of colonialism and a form of violence against nature. For Merchant, the control of nature and women was justified and enabled by a science and philosophy rooted in capitalism and patriarchy. Baconian science, for example, sexualized and feminized the control of nature, reducing women to near invisibility, and nature to machine (1983:164–82), yet the scientific revolution fitted into, and manipulated to its advantage, existing hierarchical systems of power in which the domination of women and nature had already begun (Eastlea 1981). Shiva (1988) argues that the scientific revolution caused environmental exploitation because of its association with industrialism – providing both ethical justification and technological means for the necessary exploitation of resources. Shiva fails to acknowledge that capitalist relations precipitated commodification for profit and resource exploitation prior to industrialization (Porter 1990), assuming that nineteenth-century European industrialization and modern science were mutually reinforcing, despite three centuries separating their genesis.

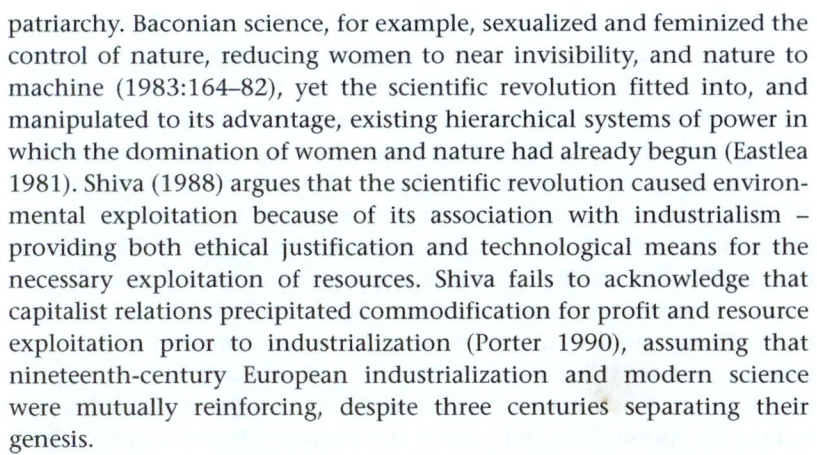

These critiques beg the question of whether ecofeminists see science itself as the problem, or the social context (white, Western, male dominated) in which science is produced. Evelyn Fox-Keller contends that science has been produced by "a particular sub-set of the human race" (1985:7), and Sandra Harding, that science is a "Western, bourgeois, masculine project" (1986:8). The characteristics of science are seen as gendered and scientific objectivity is identified as a "male" way of relating to the world, whereas intuitive ways of thinking are feminized and evaluated in the scientific paradigm as subjective (Hubbard 1990). Susan Hekman (1990) dismisses all scientific knowledge as a cornerstone of European Enlightenment thinking, and as a cultural artefact. All we know about the natural world, for Hekman, is interpreted socially, and is a form of representation. Yet Fox-Keller (1992) notes that such strong relativism is problematic because it is based on the presumption that there can be no real world "out there" to be explained.

The interface between science and culture and the physical/cultural interpenetration of boundaries of difference are problematic for ecofeminism. Donna Haraway has sought to move beyond critique of paradigms of scientific rationality as ones of domination, arguing that modernist scientific categories have constructed the separation between humans and the natural world, and to overcome human domination, we must deconstruct such categories. She contends we should conceive all objects of knowledge as sufficiently indistinct that we may speak of "compounds" of hybrid organisms (1991:212). Concepts of objectivity and objects of knowledge are constructed in terms of Western modernity, and are concepts of fixity, determinism and objectification (1988:591–6) that encourage dominant groups of humans to conceive the natural world as objects for human use. She contends we should approach the natural world not as objects but as "agents" constructed via narratives.

Haraway insists on disturbing the boundaries between human and animal and her discussion of shifting discursive regimes of gender/nature/difference in biotechnologies is always lively, interesting and helpfully provocative. However, the problem with Haraway is her (albeit inconsistent) adoption of an extreme social constructionist position that denies any reality to non-human animate beings and things, beyond human constructions of them. Feminism is at fault for deploying Enlightenment tradition in order to separate women from other animal species (Plumwood 1991:19), and Haraway's blurring of human/animal boundaries may be useful in disturbing a paradigm of human exceptionalism and human domination. However, rather than deconstructing the physical, it may be helpful to acknowledge socially mediated (but not determined) physicality for most beings. This may guard against universalizing tendencies in speaking of "nature" and discourage conceptual human exclusivity. As we saw in Chapter 3, complexity understandings of physicality, have made more nuanced attempts to understand the specific nested and interpellated qualities of life than the "leaking" of a boundary.

Haraway's dismissal of scientific knowledge as a "story" (1991:187) with no greater degree of objective "reality" than any other is highly problematic. Shiva (1988) is right to argue that those who would critique the bias of knowledge claims and claims-making processes are not required to reject an attempt to comprehend the world in rational terms, nor the idea that knowledge can be subjected to critical evaluation via empirical testing. I accept that theories and methods are shaped by prevailing discourses, and think that no form of inquiry can be utterly free from discourses of social power and contestation, a position with which as McLellan has remarked, "no modernist would disagree" (1995:402). This said, I would reject an extreme relativist position that asserts no knowledge is an improvement on any other (Haraway 1991:77). If this is the case, one wonders why Haraway bothers to write about gender or nature, or indeed anything else for that matter. As Somer Brodribb (1992) has put it "anything goes and nothing mat(t)ers." The postmodern conception of humans as self-constructed rather than (also) organic creatures is utterly modern. It is, as Mies *et al.* (1999:197) point out, a reiteration of the Enlightenment duality between nature and culture. It does not problematize or contest this duality but reproduces it. Knowledge produced from inquiry that attempts objectivity, informed by an understanding of discourses and structures of power in complex interaction may be an appropriate balance between uncritical avocation of universalizing truth claims, and the consensual stasis of extreme localized relativism.

Ecofeminism is incompatible with a strong social constructionist position. Mellor (1997) argues ecofeminism cannot accord all agency to human society, for whilst social organizations and intra-human power relations affect both how we understand the world and how the world is constructed, "the physical materiality of human life is real however it is described" (1997:7). As I argued in Chapter 2, nature, and the scientific understanding of it, may be socially constructed and interpreted by humans, but nature also has its own agency (Soper 1995). It is a form of sociological reductionism that reduces nature to culture. I have already argued that deep, social and socialist ecologists must be ontologically realist because there has to be an environment out there to be damaged, exploited and preserved. All strands of ecological thinking in many ways, stand against postmodern approaches and have often been homogenized and caricatured as biologically determinist by their postmodern critics. As feminists (even those sympathetic to postmodernism) like Diana Fuss (1988), and green social theorists like Dickens (2000), argue however, for social theorists to "confront" biology or to take biological considerations on board does not mean that they are therefore somehow essentialist. Environment and society are both biologically constituted and socially constructed. What links all the differing ecologisms is that they all adopt a form of realist epistemology. There are real environmental problems with social causes and consequences; this real world might be characterized by what Soper calls "radical uncertainty" (1995:144). The trajectory of human–environment relations is uncertain in formation and in outcome, and "risk" (Beck 1992, 1999) rather underplays our political dilemma. One of the ways of getting to grips with such uncertain futures however, is to examine more specifically the overlapping forms of domination which shape social relations with nature and affect the specific formation of our embedding and embodiment in the complexities of the web of life.

Webs, spheres and systems: inter/relations of domination

In examining the intersections of domination, and deconstructing our presumptions about women and nature, I have argued that ecofeminizm is unlikely to be essentialist. However, whether closer to radical feminism (Daly 1988) or ecology (Plumwood 1993) or socialism (Mies 1986), there is a tendency toward conflation in ecofeminist accounts, which can invite criticism for an under-theorization of difference, and therefore, of homogenization. This conflation is, in part, a question of political taste, a holistic worldview such as that of most ecofeminists, eschews the separating out, or unlinking of the processes, which result in the practices of harm to which ecofeminists stand opposed. Yet as Warren (1997:3) suggests, perhaps the crux of the ecofeminist project is "(e)stablishing the nature of the connections ... between the treatment of women, people of colour, and the underclass on one hand and the treatment of non-human nature on the other." In my view, this must involve consideration of separate processes and systematic relations of domination. If we retain some notion of system (which most ecofeminists implicitly rather than explicitly do) but do not consider a multiplicity of systemic relations, what we are left with is a rather loose theory of patriarchy, which is presumed to account for a wide range of repressive relations.

Karen Warren argues that a "patriarchal conceptual framework" is responsible for a range of oppressions (1994:181-86). This conceptual framework has five interrelated features: hierarchical (Up-Down) thinking, value-dualisms which organize reality into oppositional and exclusive pairs, "power-over" conceptions of power which maintain relations of domination and subordination, conceptions of privilege which support such power relations, and a logic of domination – an argumentative structure which justifies these relationships (Warren 1990:122-3). These five features, she contends, are simultaneously "sexist" and "naturist" and she uses a range of examples to demonstrate that the interlinked forms of gendered and natured domination are manifest in material reality (Warren 1992). This patriarchal conceptual framework equates women with nature, and an anti-nature (Western) culture also feminizes the natural environment. Warren further argues that the feminist project must, given the global diversity that is/are women, be a movement to end the whole plethora of oppressive "isms": racism, classism, heterosexism, anti-Semitism, agism, eurocentrism (1997:4, 1987:133). Ecofeminism is distinguished by its "insistence that non-human nature and naturism" (i.e., the unjustified domination of nature) are feminist issues. Due to the intermingling of such varied influences, the "quilt" that is ecofeminism (Fox 1997:155) has a unique ontology for the articulation of various dominations. Warren suggests we might capture the theoretical parameters of ecofeminism by three spheres of "feminism," "native/indigenous, local perspectives" and "science, development and technology." Ecofeminist theory is seen to take place in the multiple overlapping center (1997:4). However, her spheres are not commensurate for she does not distinguish activism from theory, and some spheres remain undefined and ambiguous. What is disappointing is that she does not build upon the "conceptual framework" of interlocking dominations that was suggested by her earlier writings and remains in need of detailed theoretical exploration and explanation with respect to each specific formation of domination.

Traversing similar terrain but more successful, is Val Plumwood's understanding of different forms of social domination as intimately connected. She sees gender, nature, race, colonialism and class as interfacing in a "network" of oppressive "dualisms" (1993:2). They exist as separate (autonomous) entities but are also mutually reinforcing in a "web" of complex relations (1993:194). Systems of oppression based on difference are interconnected and form an interlocking "web" (1994:78–9). This does not mean different forms of oppression are indistinguishable; they are relatively autonomous, distinct yet related. Although Plumwood argues oppressions within the web have "distinct foci and strands" and "some independent movement," she adopts a similarly conflationary approach to Warren in arguing "ultimately," forms of domination have "a unified overall mode of operation, forming a *single system*" (1994:79, my emphasis).

In attempting to draw together analyses of intra-human hierarchy with the domination of nature, Plumwood (1993) argues "reason" is a master narrative of Western culture (1993:196), which is key to the construction and maintenance of oppressive relations around gender, nature, race and class. Western social thought is based on dualistic concepts: culture/nature, male/female, mind/body, master/slave, civilized/primitive, production/reproduction, public/private, human/nature (1993:43). This constructs difference in terms of power relations of domination and subordination, and Plumwood's solution is to replace dualist concepts with a non-hierarchical concept of "difference" (1993:59). We need to integrate differences in society and not see them in hierarchical terms. However, despite all its promise, Plumwood's position is not so far from the postmodern theories of which she is so critical. Her analysis of the way dualism operates emphasizes the power of ideas, rather than of discourses concretized in practices and material relations. We dissolve the dualisms by re-conceptualizing them as non-hierarchical difference. Plumwood ends up with a focus on individual change, avoids talking of social domination, and reduces human domination of the environment to an "identity" in which we are complicit.

In addition, whilst she argues oppression must be conceptualized in terms of multiplicity, in reducing the web of oppressions to "a single system," with a "common structure and ideology" (1993:81), she provides an analysis that stresses symbiosis and denies conflict.

More recently, Plumwood has moved from a model of interlocking webs of dualistic conceptual frameworks, to accepting the potential autonomy of systems of social domination. She is now keen to defend and retain a concept of anthropocentrism as a powerful analytic and political concept key to understanding human relations to the natural environment. Outlining the problems inherent in a "cosmic" concept of anthropocentrism, as articulated by most deep ecologists (1997:331), Plumwood brings together the insights of a range of -isms in order to understand the interlocking of domination(s), and ground anthropocentrism in a way that specifies different kinds of human domination of nature. Plumwood now argues for the existence of a "common centric structure" (1997:336, see also Hartsock 1990). This structure places an "omnipotent" subject at the center and constructs non-subjects as having various negative qualities, and underlies racism, sexism, colonialism and "naturism." The features of this centric structure are: the exclusion of Otherized groups, the homogenization of differences within both dominant and Otherized groups, denial (of dependency on the Other), incorporation (where the Otherized are judged negatively, to be lacking) and instrumentalization (Other's become a means to an end of the subject group). The interlinked centric model moves us away from a single system framework, but difficulties remain. In her three examples (Eurocentrism, androcentrism and anthropocentrism), two questions emerge. First is the question as to whether the rather vaguely defined centrisms are actually systems of dominatory relations. As they stand, Plumwood still supposes they are ideological frameworks, rather than discursive regimes embedded in institutions, processes and practices and we have an account that does not embed the centrisms in materiality. Second, the elements of the centrisms are parallel. I am not convinced that we might structure a multiple model, systemic, or centric or otherwise, in quite this way. The elements, or as I prefer, structures (specific institutional formations and practices) of the "centrisms" may not be identical given their specific patterns of historical and geographical emergence and the kind of difference they primarily construct. The first two chapters of this book suggest this much, and I will attempt to demonstrate the non-parallel qualities of systems of domination in Chapter 7.

In unpicking the intermeshing of differences within an embodied ontology of *structured* domination, Maria Mies analyzes the exploitation of women and of nature in terms of social structures of domination: capitalism, patriarchy, colonialism, militarism and the state (1993:223–6). Although inevitably somewhat dated, her impressive *Patriarchy and Accumulation on a World Scale* (Mies 1986) is an attempt to integrate feminist and class analysis within a framework of anti-colonialism. Global patriarchal and capitalist accumulation "constitute the structural and ideological framework" within which we might theorize gender relations in all their diversity (1986:3). Mies suggests that the gendered division of labor is at the core of the linked exploitations of women by men, southern countries by the wealthy northern states of the globe, and the natural environment by human society. Mies traces the historically linked processes of colonialization and what she calls "housewifization." The current distinction between production and reproduction has devalued women's "essential" unpaid domestic labor (1986:53), and ensured women are relatively disadvantaged in the job market. The flexibilization of the workforce in the West is gendered in inception and effects, with women segregated into the more vulnerable and poorly paid sectors of the job market. Similarly, the domestication of women in poorer regions of the globe means that their labor can be bought more cheaply than that of men and they are thus the "optimal labour force" for transnational corporations (1986:116–17). Mies sees women of the first world as differentially implicated (given divisions of class, status and wealth) in what Andre Gorz (1989) has called "compensatory consumption" (Mies 1993:251–9).

Mies deploys a theory of capitalist-patriarchy and draws upon various theories of development/underdevelopment. Capitalism and patriarchy are mutually dependent and constitutive being "one intrinsically interconnecting system" (1989:38). Whilst capitalism and patriarchy are fused into one system, they crucially interact with two other oppressive formations: human relations with nature, and relations between developed and underdeveloped societies. Different specific relations of domination can be seen at different historical junctures, in Mies's account, depending on the way in which the interests of socially dominant groups coalesce (1986:87–8). What is undertheorized in her account however, are systemic human relations with nature, which seem an aspect of capital accumulation rather than having specific links to gendered relations of power.

In presenting her years of collaborative work with Veronika Bennholdt-Thomsen and Claudia von Werlhof, Mies ecofeminist credentials can be more clearly seen. Their "Iceberg model" of Capitalist Patriarchal Economics draws together dominations of nature, class, gender and region. According to this model, only certain aspects of the economy of exploitation are visible – that of capital and wage labor. What is invisible is nature, treated and exploited as a free good in contemporary global corporate practices. In addition, the labor of children, peasants in a subsistence economy, homeworkers and housewives is similarly invisibilized (1999:31–5). Economic growth, or capital accumulation, in their view, continues due to the maintenance of colonies and the seeking out of new areas for colonization (such as the bodies of plants and animals, Mies and Shiva 1993). Mies *et al.*, propose a radical solution to social injustice and environmental destruction: a "subsistence perspective," where subsistence production is prioritized and productive work is neither exclusively public nor linked to the money economy. The gendered division of labor from the local and regional to global formations will be radically undermined, and stewardship will necessarily replace exploitative attitudes toward nature. However, whilst Mies has much to say of gender, and offers important insights on development and ecology, her explanations revolve around a theory of capitalism. The dominations of colonialism, of gender, and of nature appear rather as systemic effects of capitalism, and although she makes constant use of the term patriarchy, she does not conceptually deploy it as she might.

Coming closest to a multiple systemic account is Mary Mellor (1992) who argues that we need to integrate an analysis of different formations of social domination into an analysis of human relations with the environment. Mellor attempts to address the limits of Marx's understanding of capitalism and draws on both Engles and the arguments of some eco-socialists in arguing that an adequate theory of capitalism must acknowledge the concepts of natural capital and natural limits (to economic growth). When we consider such natural boundaries, capitalism is incapable of meeting the basics needs of the "world community." Mellor sees patriarchy as a structured social system of gender difference and domination has problematically divided the human world into feminized private and masculinized public spheres, "placing on women the major responsibility for nurturing and caring values and activities" (1992:251). An important distinction in Mellor's account is that this is an "imposed altruism" and does not necessarily foster a particular relationship to the environment by extension of empathy beyond for example, care of children and other family members. Whilst the public face of capitalist production takes as its premise an autonomous individual, the nurturing and supportive private world remains invisible. To focus on women's work enables a critical analysis of "the construction of a social world which has its material base in women's time and work" (1997:174).

In knitting together the structuring of gender and of capitalism, Mellor brings time into the equation. Women's imposed altruism means that they are not so fully subsumed within advanced capitalist time as the majority of men, but are more closely influenced by biological time or as Jeremy Rifkin (1987) refers to it, "ecological time." Biological time is Mellor's corollary to Mies *et al.*'s "subsistence perspective" but it is not so radical in that it cedes to capitalism the benefits of (albeit reduced) economic growth. The problem of speed is seen as key to contemporary formations of capital and gender relations, "by separating production from 'reproduction' and nature capitalist and socialist patriarchy has created a sphere of false freedom that ignores biological and ecological parameters" (1997:173) and this is a sphere in which only the fictive economic man might flourish. Whilst Mellor acknowledges that women have differential relationships to, and responsibilities for, the "work that supports biological time" (1997:175), she remains rightly insistent that this work is gendered, and that it throws into sharp relief the dilemmas we face as an embodied species. The material relations of women's lives are not captured by theories of capitalist relations, as women's lives are crucially located at the boundaries between the public and private, society and nature.

Mellor prefers "filiarchy" – the rule of irresponsible sons, to describe the systematic domination of both women and the "planet" (1997:193). This notion is not substantiated nor argued for in any sustained way, and appears almost as an afterthought. I do not think Mellor is really suggesting anything more than a renaming of theories of patriarchy here, and I am not convinced this adds to her analysis. What is interesting about her brief terminological discussion however, is that, like so many of the Marxist, socialist, feminist and ecofeminist accounts she incisively critiques, she falls into the conflationary trap she seeks to avoid, and states that filiarchy is a system of gendered and natured domination. I think Mellor has more to say on the interrelations between patriarchy, capitalism and the domination of nature, concerned as she is, with a matrix of interlocking and overlapping systemic dominations of "race" and post/colonialism, sex/gender and capitalism. Key to this analysis is the sex/gender division of labor around human embodiment, wherein women bear much of the burden, in their bodies and their lives, for our embodied ontology. I want to examine in greater detail the specific formations of embodiment and human–nature relations to be found in re/productive labor, and this is the focus of Chapter 6.

My criticism of Mellor is that she backgrounds the human domination of the environment in her analysis of capitalism and gender relations. Her analysis is both wide-ranging and complex, but despite being so, in mapping the connections, the differences that do not overlap are left unexplained. Different systemic relations of domination make themselves felt in varied formations and to different degrees, and these relations are spatially and historically located. Chapters 3 and 4 made the case for systems of anthroparchy and patriarchy, which, whilst interlinked and overlapping, are distinct from each other and from all other systems of difference-in-domination.

This chapter has indicated many of the connections already established between gender and nature. Some feminists have contended that women's material life experiences render them different by virtue of their sex – a more embodied material experience which may potentially encourage a concern with practices of harm against nature and which, under contemporary relations of social domination, means that their embodied experience is naturalized. Others have deconstructed the association of women and nature in patriarchal discourse, some arguing this association was a result of modernity and the gendered discursive regime(s) of scientific rationality. Some ecofeminists consider such an association between women and nature to be the product of far earlier paradigm shifts, and that the domination of women and the environment preceded the transitions to modernity in the West, whilst it was also intensified and assumed different forms with modernization.

Systems of domination based on difference interrelate in complex and contradictory ways, and are best conceived as distinct, yet interdependent in complex and specific ways. Rather than the formulaic overlapping of spheres, or a congruent "web," systems of domination interrelate and intersect in the more complex manner of the planes of a snowflake, which assumes *unique form within a particular time and spatial location*. Thus particular instances of oppression demonstrate unique and specific articulations of various combinations of domination. The difficulty lies in how interconnections between systems of oppression may be investigated, and the strength of multiple systemic analyses is the potential to investigate the interconnections between autonomous yet related systems of oppression.

By eschewing complex systemic approaches, ecofeminists have provided rich accounts which remain open to the critique of reductionism by either seeing the oppression of both women and nature as a product of patriarchy or (mainly) of capitalism or an overarching logic or system of a rather unspecified domination. The tendency to reduce the domination of the environment to ill defined patriarchal and/or capitalist structures and processes, and to identify patriarchal discourses as both gendered and natured has made ecofeminism vulnerable. Difference has been unwittingly marginalized when ecofeminism is so well placed to capture the complexity and power dynamics of its dominatory formations. Such a conflationary approach carries a danger of reductionism in which patriarchy, in the main, is seen as an explanatory theoretical schema for other kinds of domination. There are sufficient differences in the form and degree of domination to necessitate a multiple systems approach to the examination of the relationship between gender and ecology. This allows us to capture the complexities of structural and systemic dynamics without marginalizing difference. Chapter 6 considers how the multiplicities of social domination can be seen as embodied structures of gendered and natured power, whilst the final chapter draws together the elements of a multiple systems approach.

6
Embodiment, Materiality and Symbolic Regimes

> ... even though we share placement in the phylum of vertebrates, we inhabit not just different genera and divergent families, but altogether different orders ...
>
> ... How would we sort things out? Canid, hominid; pet, professor; bitch, woman; animal, human; athlete, handler. One of us has a microchip injected under her skin for identification; the other has a photo ID California drivers license. One of us has a written record of her ancestors for twenty generations; one of us does not even know her great grandparents names. One of us, product of a vast genetic mixture is called "pure-bred." One of us, equally the product of a vast mixture, is called "white" ...
>
> ... We are, constitutively, companion species. We make each other up, in the flesh. Significantly other to each other; in specific difference, we signify in the flesh a nasty developmental infection called love.
>
> Donna Haraway, *The Companion Species Manifesto* (2003:1–3)

The notion of corporeality and a concern with the embodied qualities of gendered power has been key to the development of Western feminist theory. The defining of birth, mothering, reproductive management, the praxis of sex, domestic violence and adornment as questions of politics, cultural representation and social organization, has meant that feminism has often been embodied theory. Ecofeminists have emphasized embodied relations of power with respect to reproductive technology, foodways, and specific environmental impacts on women's bodies and health. Elements of other ecologisms might be read as embodied theory. The "wilderness reverence" of deep ecology is both implicitly suggestive and explicitly demonstrative of the engagements of bodies with/in their environments. As we have seen, there is an ecofeminist critique of such literature as encapsulating an exclusionary experience of natured embodiment in its emphasis on difficult and demanding physical experiences. There is also a postmodernist critique which focuses on the "purist" elements of such endeavors, which often commend some form of "return" to nature and a holism which posits an "essentialist" relation between humanity and "environment." Nevertheless, a strong element of thinking through the body is to be found in such writing.

This chapter begins with a consideration of embodiment with respect to social structures of multiple differences. Foucault's work on the body implies the embodiment of structure, and feminists have sought to articulate how social difference co-constructs gendered physical bodies. I move beyond Foucault in considering the ways corporeality, both human and animal, are co-constructed, and examining how material practices and corporeal entities are shaped and shifted by discursive or symbolic regimes. It stretches the definitional boundaries of "corporeal" to include vegetal bodies or "land/earth" masses, and perhaps "material" is more appropriate a descriptor for their consideration. Rosi Braidotti (2002:2–3) suggests that we live in a "permanent process" of hybridization, inscribed through individual and social bodies and professes an enticingly named "enfleshed materialism." Yet there are some serious problems with her postmodern analytics of the body, which turn on questions of essentialism/non essentialism that have percolated so many of the pages of this book.

Incorporating insights from feminisms and ecologisms, I argue that bodies themselves are best conceptualized as systems (with both "open" and "closed" properties) within which other systems may be found, and which are "nested" within other physical and social systems, few of which are discreet. I consider the ways gender, nature and other differences-in-domination play themselves out in material practices and discursive regimes of culture. This involves a consideration of the dynamic of human, domestic animal and agricultural plant reproduction in the context of anthroparchal and patriarchal relations in late modernity. Feminism has taken account of increased potential for medical intervention in fertility, pregnancy, and childbirth, and the development of new reproductive technologies. Some feminists voice concern as to potentially draconian implications of such technologies for women, whilst others see particular benefits. New reproductive technologies are best understood through the framework of their development as profit enhancing tools. Their development as a gendered form of reproductive management, which impinges on human bodies in specific ways, is derivative of their legacy in the management of non-human domestic animals.

The chapter closes with a consideration of embodiment in a world that may increasingly decenter the human. One of the best-known mothers of the posthuman, Donna Haraway, has influenced a wide range of scholars with her popular cyborg motif which functions as both a future imaginary and as a descriptive category for contemporary fusions of techno-social/nature. In the quote from Haraway with which this chapter begins, we can see both the power of analogy and the challenge of difference. I argue a need to think across the difference of species for an embodied and enfleshed posthuman condition, which may be more aware of both the potentialities and boundaries of the embodiment of multiplicitous species.

Embodied theory

Following Foucault, theorists of the social body consider physical bodies as unfinished resources (Shilling 1993:103) that need to be "trained, manipulated, cajoled, coaxed, organized and in general disciplined" (Turner 1992:15). Iris Marion Young (1990) has argued that "embodied" theory takes the body seriously as a physical entity, examining the embedding of gender in bodily postures, musculature and texture. A structural reading of Foucault brings embodied theory into a complex modernist frame, but can only help us understand the exterior body. We need a critical realist approach in order to understand the materiality of human embodiment and the body as "an irreducibly physical phenomenon engaged in a dynamic relationship with its social surroundings" (Shilling 2003:x). Postmodern theories of bodies, as essentially both social and plastic, are reductionist. I suggest that the systemic approaches of complexity theory, which eschew teleology and fixity, and emphasize the body as a network of complex interrelated social/bio systems of emergence and becoming, can overcome the culture–nature divide which has preoccupied social theorization of our embodied and ecologically embedded condition.

Liquid flesh: bodies without organs

Much recent feminist work on the body and embodiment has been poststructuralist, and has tended to render corporeality entirely social. As Rosi Braidotti surmises, the body is the:

> Complex interplay of highly constructed social and symbolic forces: it is not an essence, *let alone a biological substance,* but a play of forces,

a surface of intensities, pure simulacra without originals. (2002:21, my emphasis)

This is a very strange "enfleshed materialism," for the body is without fat and muscle, blood and bone, and such theorizing draws on Giles Deleuze's notion of the "body without organs" in which bodies are collections of symbols. Braidotti's symbolic body is a receptacle of the (fragmented, plural) "self" and the "subject" (Gatens 1996), and is heavily influenced by Luce Irigaray's (1977, 1996) work on difference and feminine sexuality. As we have already seen, the matter, in Judith Butler's *Bodies that Matter* constantly slips away from her analysis (often on her own admission, Butler 1993:68). Butler (1990:136) defines the body as having "no ontological status apart from the various acts which constitute its reality" which fails utterly to understand the specific ways in which we are both an embodied species and one embedded in relationships with complex "naturecultures" (Haraway 2003:100).

Elizabeth Grosz (1995) also explores the body deploying a Deleuzian analytic and emphasizes the fluid qualities of bodies as inscripted and dynamic surfaces. The effect of this is that the body becomes an extended surface, almost conflated with the skin. Such social constructionist theorizations are prey to their own reductionism and essentialism (Birke 1999:138) in that they exclude physicality, thereby fixing it and assuming it is constant beneath its cultural inscription. Whilst concepts of the body as a location on which culture imprints itself may tell us something of such cultural normativity, it tells us little of corporeality, particularly that of interior bodies within the skin. In addition, the pluralist analysis is less helpful in ascertaining the impact of social difference on the body than the more clearly structural approaches of Bourdieu, and in particular, Foucault.

Structuring the body

Foucault is interested in the differentiated and "eccentric" (Lopez and Scott 2000:83) aspects of social structures, particularly those in which complex networks of power relations impinge on the physical body. Discourses are conceptualized as compatible with systemic and structural notions wherein:

> Systems of power bring forth different types of knowledge, which in turn produce material effects on the bodies of social agents that serve to reinforce the original power formation. (McNay 1994:63)

Here, discourse becomes a structuring principle which governs beliefs and practices, and this conflation is most clearly seen in Foucault's writings on government, where he rejects a legalistic notion of the powers of the state, arguing that "government" is a process of disciplining a population through "biopolitical" control (Foucault 1976b:42). Discourse is a structuring mechanism, creating and maintaining social institutions that through their disciplinary procedures create a "society of normalization." Foucault is primarily concerned with disciplinary discourses that operate via the body, and are relations of power that marginalize deviancy (Said 1988). Foucault refers to particular cases of the construction of madness (1971), crime (1979) and sexuality (1981) and the discourses of power operating within and through the asylum, the prison and the professions of psychoanalysis and psychotherapy to "normalize" the embodied subject. Problematically however, for Foucault, gender and sexuality are arenas wherein the individual has a fairly substantial "practice of liberty" in the interpretation of discourse (1985:10–23). The emphasis in the first volume of *The History of Sexuality* for example, is on the diffuse, heterogeneous and changeable nature of power relations, and thus the Victorian family is not seen as a repressive force, but an arena in which diverse discourses of sexuality are played out (1981:100). The historical processes of the gendering (Laquer 1990) and naturing (Jordanova 1989) of bodies in modernity is lost in Foucault's account.

This said, feminist theorizations of the body have often drawn very usefully, on Focauldian analysis in outlining the powerful array of disciplinary practices, which produce "docile bodies" (Foucault 1979:138) through external control and internalized belief. Such work has involved an analytics of bodily behavior, such as movements, gestures and adornments, and modifications such as dieting, which can be seen as patriarchal forms of embodiment (Bartky 1997), arising at intersections of relational, institutional and embodied structures of systemic domination. Bodies are sites of gendered social control (Bordo 1997:90) the mechanisms of which have become increasingly internalized as they have become subtle (e.g., as women's bodies in "modern" states are no longer male property, Weitz 2003:5, hooks 1992).

A further matter for concern has been the peculiarly "feminine" modalities of embodiment about which feminist theorists of embodiment have been concerned. Pierre Bourdieu (1984:466–7) speaks of values and norms incorporated into the body as dispositions to act (ways of talking, standing and walking), as "embodied social structures" which are embedded in social inequalities. Deploying such notions, Young (1990) has described the ways in which gender norms suppress the physical potential of women, whose bodies demonstrate comportment in a restricted spatiality. For Sandra Bartky (1982), the invisible barriers on physical and erotic expression, and psychological pressures of gender normativity cause women not only to adopt particular ornamentation, gestures and appearances, but also to be "alienated" from their own bodies and from female bodies collectively. Naomi Wolf (1990) has poignantly documented the effects of the "myth" of feminine beauty in America as what can be seen as a "disciplinary practice" of voluntary starvation (also Chernin 1981) and the impacts of surgical "improvement" for defective bodies. The policing of bodies may be less severe, with constant but limited "dieting," dress, "face care" and make up, shaving and plucking which make significant demands on women's time and energy (Bartky 1988). Thirty years of feminist activism has done little to disturb the Western cultural norm that the undisciplined female body is unacceptable (Barbre 2003). In Chapter 4, I argued that Walby (1990) had less to say than she might about the embodied structures of patriarchy, although she considers sexuality and domestic violence to be important components of a multidimensional understanding of patriarchy as a system of social relations, and brings together structural embodiment with analysis of relational and institutional aspects of social structure. This is in line with the foucauldian concentration upon the relations which exist between the "body and the effects of power on it" (Foucault 1980:58), and Bourdieu's concern with the ability of dominant social groupings to define their bodies as superior. Thus, we have the woman made body, in which "the feminine body-subject – is constructed," premised on what Bartky (1988) refers to as "systemically duplicitous practices." The structures that are embodied include the cultural requisites reflecting racial, age, size and proportion, shaped by systemic requisites of gender, race and age.

Such systemically duplicitous practices may be reflections of "disgust" of women's bodies, particularly the aging body (Greer 1999) that is internalized almost unmodified by women as ambivalence and distain (Lee 2003:84–90). Other practices may position the gendered body as a site of resistance however (Tolman 2003), and exemplify both internalization of normative gendered practice and the assertion of positive difference (Weitz 2001). There have been serious debates within feminism about certain forms of behavioral practice in which women "choose" to discipline their bodies through patriarchal self-hatred in order to achieve a certain appearance of "breast" (Young 2003:152–4), or facial features which are white, Western and Anglo-Saxon (Morgan 1991,

Kaw 2003). Whilst in wealthy America, cosmetic surgery is becoming a normative disciplinary practice, it is also a gendered practice cross cut by "race" and nation and available to those who can afford it. As Rose Weitz (2001:685) remarks, the female body is problematic as a site of resistance to gendered normativity, because whilst individual acts of rebellion may improve women's lives, the impact they may sustain are delimited and constrained. Dominations of difference are expressed as a style of the flesh, an enfleshed constitution of the structuring of relations of domination. This should not be seen as cultural "inscription" on the body, but what Katherine Hayles (1999:198) refers to as "incorporating practices," wherein our practice of being cannot be separated from its physical medium.

The institutions of popular culture, such as women's magazines and adverts for beauty products are tied to a multi-billion beauty industry, an adjunct of petro-chemical corporate power. In addition, international trade is also concerned with circuits of bodies, bodily representations and body servicing (Phizaclea 1990). The significance of such culturally mediated practices, despite their diffuse nature, is that women embed the gendered body in their "private" spaces. Foucault offers an understanding of human bodies as more than surfaces for inscription, but as the body is constituted by and through discourse, it does become a malleable product in its social construction. As Shilling observes (2003:70–1) this renders the biological body simply a manifestation of the social. Much feminist theory of the body can be seen as human "exceptionalist" (Benton 1992) where human cultural and social capacities are held to displace the importance of biological mechanisms in the constitution of human subjects. As Shilling argues, we need to conceptualize the body as both "simultaneously biological and social" and as "shaped by but irreducible to contemporary social relations and structures" (2003:182).

Humans, animals and embodied difference

Foucault is concerned with both the implication of individual bodies in the procedures of social institutions, and also the ways "social bodies" are controlled, organized and distributed in social and physical space (Turner 1996:161). The operation of bio-power centers first on the body, primarily conceptualized within the medical model (Foucault 1981:139), and second, it focuses on the species body, in terms of population control policies for example, and conditions of health and variance in longevity (1981:140). This goes some way to opening up structural analysis to the physical body. Further to such an end, Bryan Turner

(1992) attempts to combine an analysis of the body as a biological organism with an analysis of bodies within regimes of representation. Animals, he argues, have specialized and directed instincts, which determine their being in the world in terms of existence and survival within their environment. The human world is comparatively open, as we do not have such a specialized innate instinctual structure and must shape our world in order to survive. Humans are in an unfinished state of embodiment and must work on their environment in order to survive. This encourages them to develop particular orientations toward their bodies, and symbolic regimes through which they are represented. Such analysis cannot account for the incredible difference in the bodies of non-human animals. Neither does it account for relations of intelligence, sociality and bodily form. Animal bodies are not "given" but modified after birth through for example, agricultural practices again, of an enormous variety of forms and levels of physical intervention. This view of humans as bodily distinct belies the extent to which human societies are embedded in environments, including webs of vastly differing kinds of relation with a multiplicity of species. I would strongly dispute the notion that our *distinctive* species relation to our bodies is based on evolutionary processes that incorporate both social and biological factors (see Benton 1991), because this is the case for a wide range of animal species.

The human body is not infinitely malleable, nor are other animal species simply set in a fixed physicality. Bodies are to some extent socially constituted, but are also absolutely material phenomenon and "medium" for the very constitution of society (Shilling 2003:209). Such insights apply to animal bodies interpolated strongly in human social and economic networks, institutions and practices. Anthropoarchal structures and disciplinary practices can be strongly evidenced in the bodies of humans and animals, and such practices are socially constituted through conceptions of nature and gender, and discursively often also racialized. Structures of power which impinge on some human and some "animal" bodies similarly impact on the experience of embodiment, although in the case of domestic cattle and companion animals for example, such experiences are far more challenging for we humans to comprehend.

Ecologisms have theorized the body implicitly for the most part, rather than explicitly. Whilst deep ecologists have had much to say about our "ecological self-hood," they have been reticent in speaking of our embodied condition. Yet it is within the conception of an expanded self that we are encouraged to "experience ... oneness in diversity"

136 *Developing Ecofeminist Theory*

(Naess 1985:261), and this is a question of embodied cognition. As Kirkpatrick Sale (1985) advises, we come to understand our own bioregion through physical proximity, or as Bill Devall puts it: "The boundaries of self-identification are those we walk through" (1990:60). We discover our environment and the other species with which we co-habit or "dwell" by physically placing ourselves in the space. Certainly, in the deep ecological literature, experiencing our dwelling is prescriptively un-urban and often anti-urban, but wilderness utopianism aside, it encourages an understanding of ecological embeddedness. This notion of physical embedding in environmental systems, and of a corporeal sense of "place" captures certain kinds of embodied difference whilst being blind to others. Social difference and disadvantage is absent, but the difference and diversity of non-human biota is fore grounded. The purist conception of nature that critics associate with such thinking is not always evidenced (see Devall 1990:160), but I consider that deep ecologists go too far in abstracting the sensing body away from social difference. The notion of embodied humans in mixed communities of nature cultures captured in some accounts however, is an important corrective to the generally disembodied condition of social and political theory.

For Lynda Birke (1999:1–2) the feminist critique of biological determinism has led to a marginalization or even phobia (Spelman 1990:127) of physical bodies, particularly the materiality of the internal body (with exceptions, Stacey 1997). Emily Martin (1996:106) has pointed out that despite the reductionism and mechanism apparent in genetics, there is a cultural shift, reflected in biological theory, to interpreting bodies in non-mechanistic ways. Birke's own model is a further development of such trends, and she draws, albeit not always explicitly, on systems and complexity theory. Bodies might be conceptualized as located in a live and dynamic environment with which they co-evolve, changing evolutionary niches rather than simply "filling them." In this context, we might consider the widely divergent forms and qualities (both internal and external) that bodies assume across the globe, which question prevailing notions of fixity and constancy in speaking of the embodied human condition. Social change and the impact of war, famine and cultures of consumption affect bodies physically. Physiological systems are embedded in the material body in terms of "readouts from 'inscription devices', which themselves embody previous technique and expertise, and come to stand for the functioning of the biological material (or animal body) itself" (Birke 1999:107).

Thus human/animal bodies can be seen to embody layers of complex social relations and practices, and in turn, physiological systems are themselves embedded and implicated in wider bio/social systems. Birke notes the extent to which understandings of complexity are permeating biological theory, using the example of the heart. Despite the mechanistic notion of a body composed of replaceable pieces of machinery exchanged through market relations (Kimbrell 1993), and across species boundaries (with zenotransplantation, of pig organs in particular), there is an increasingly influential discourse of electrical chaos. This chaos is not random however, but is persistent instability, which is at the same time, self-regulating. Our bodies, organs and processes in complexity thinking, are dynamic and self-organizing with tissues and cells constantly in change and flux. Bodies are not genetic blueprints in this conception, but are organisms with internally active and co-operative (symbiogenic) processes generating the external form. Whilst, as Kauffman (1995) suggests, we may appear "fully" formed in our human embodiment, our bodies are not fixed in this sense. Rather, they are complex and elegant bundles of possibility and becoming, which are dynamically engaged with the physical and social systems of the environment.

In order to further examine how structures of gender, nature and other domination of difference are embedded and embodied, I want to look at the materiality of production and reproduction. Symbolic regimes and material practices of gender/nature implicate human and non-human bodies. I want to extend the notion of bio-power in discussing the regulation and control of "life" that can be seen so clearly in examining re/production. The gendering of medical control of women's bodies has long preoccupied feminists (Ehrenreich and English 1973, Oakley 1976, 1980, Riessman 1983), but debates on the social construction of "nature" and embodiment are thrown into sharp relief in considering technological reproduction, and this is perhaps where feminist/ecofeminist theorizations of embodiment are most advanced.

Alienated re/production

There are specific areas of overlap between embodied structures of patriarchal fertility and reproductive management, and anthroparchal structures of fertility and reproductive management within farming, which are manifested through the bodies of agricultural animals. Feminists have long argued that the experiences of pregnancy, birth and childcare are often alienating for women. The work of reproduction is privatized and excluded from the valuation and renumeration relating to paid employment, and such alienation is linked to continued public/private duality. Ecologism has problematized the specific ways modernity has impacted on "nature" through the invasive modifications of industrial agriculture within an analytics of production, rather than re/production. The very modifications and technologising of agricultural space however, has meant it is a nexus of hybridization (of non-human species) and a key site within which the co-construction of the natural and artifactual is played out. This is not a benign game, but one in which the dynamic but donimatory intermeshings of power relating to gender, nature, race and class are very much present.

Some radical and ecological feminists have drawn parallels between the control of animal sexuality and reproduction and the application of reproductive technology to women. Some argue the reproductive experience of women in Western high modernity increasingly resembles that of "meat" animals. I do not accept the pessimism of such accounts, which, in their often exclusive focus on the patriarchal significance of reproductive control, are nature blind. Others argue that the uncertainties of human conception in an age of in vitro fertilization and embryo transfer are reflective of the postmodern condition in which biology has a non-essential relation to the operation of "real" bodies. However, radical, ecological and postmodern feminists all underestimate the embedded ontologies of techno-nature, and tend to ignore the applications of such technologies across the centuries of modernity and across the species divide. For thousands of years, human agriculturalists have co-constructed and hybridized other species. There is often the presumption that the development of reproductive technologies has profound implications, but such manipulation of fertility and fecundity is different in degree and specific formation, and is not a paradigmatic shift.

Gendered production within the social relations of capitalist nature

Feminism has problematized the apparent dichotomy between a public sphere of productive labor and the privatized work of women as reproducers, nurturers and socialisers of the species "human." Drawing on analogies between Marxist notions of "modes of production" and conceptions of "modes of (biological) reproduction" (Firestone 1988), some have suggested that these processes are not dichotomous (Bartky 1976), and that social relations change significantly with shifts in technologies of reproduction, in terms of contraceptive choices, fertility possibilities, shifting technologies in the workplace. The bodies of humans, other animals and plants have been the subject of human "work" and technological interference prior to the advent of "modernity" in Europe, albeit there was likely to be less cultural distinction between "environment" and "society" (Ingold 1994). However, there are particular ways in which capitalism and other formations of social organization's and power relations, cast the relations between embodied humans and the environment in which we are embedded.

For Marx of course, alienation under capitalism involves the alienation of labor. Labor distinguishes humans from other animal species or the free, conscious and creative transformation of nature, in line with human needs. Human labor is objectified in its products, which are appropriated by capital and become part of an alien force. Engles was particularly concerned with the impact of the division of labor under capitalism, which made working bodies fit only for limited and repetitive activities. Marx and Engles were also concerned with the corporeal conditions of consciousness, and argued that human development took place as a result of a dialectical relation between nature and human abilities to transform the conditions of life through labor. Peter Dickens (1996) expands on these important insights through a detailed analysis of the division of labor in contemporary capitalism, which structures the ways we work on nature to produce the things we need. Organizing labor in capitalist societies around the production of marketable goods on the basis of increased production and consumption to satisfy the profit motive means that the natural environment is exploited. Nature is transformed into objects valued only as property or commodities under capitalist relations. The more sophisticated our ecologies for humanizing nature become, the more alienated we are from nature. Dickens considers that all humans, animals and inorganic entities possess latent powers or ways of being and acting, which combine in complex ways in certain circumstances (1996:10). Thus there are likely, given the complexity of interlocking social hierarchies, to be "multiple and combined forms" of alienation which affect orientations to our embodiment and our ecological embedding.

Dickens draws on Marx's notion of the "circuit of capital" in order to explain the social process by which we "combine(s) with nature to produce commodities" (1996:43). Technology and labor power combine with "raw materials taken from nature" in a labor process to produce marketable commodities. In addition, the powers of nature are utilized as production processes, for example, in the production of "animal products" such as milk and eggs. Biotechnologies are deployed as adjuncts of capitalist process. Reproductive technologies and genetic engineering illustrate the applicability of this "multi-dimensional" notion of the alienation from nature of both humans and certain other species. Here, the materials incorporated into the production processes are not inert, but living beings with causal powers and properties. In both cases, body parts and elements are separated from their context and manipulated to produce new hybrid entities. Thus for Dickens, animals "species and natural being" (1996:62) is exploited through manipulation and distortion, as a commodity, through intervention in the ways in which animals reproduce. The powers of animals, similarly to industrialized workers have been subjected to progressively more intensive rationalization and accompanying automation and mechanization in order to produce commodities of meat and milk for human consumption. Dickens argues that animal lives are subjected to an increasing division of labor (through intensive production) and at the same time, loose their own capacities (to seek out their own food for example) (1996:63). Thus the "species being" and "natural being" of animals are being treated as disaggregated wholes, parts of which some humans deal with (feeding, milking), and parts of which many humans consume (as meat and derivatives, eggs and milk). The effects of such treatment leads animals to be alienated from themselves, other members of their species group, and their own offspring.

The development of human agriculture can be seen as a mechanics of humanizing nature, but both animals and plants have undergone more invasive technological manipulation in the last 30 years, as genetic engineering is increasingly the norm in "enabling" nature to transcend biological limits and become more productive within a highly industrialized frame. Dickens (1996:114–5) notes that capital is increasingly intervening through biotechnology, to regulate the reproduction of seeds. In the "South," traditional praxis has meant that farmers can save seed after harvesting and exchange locally to ensure genetic diversity (Shiva 1998). Bio-diversity prospecting in poorer countries, argues Shiva (1993), means that the new colonialism of biotechnological transnational capital can patent plant life forms and extend commodification into new areas of life. In addition, the manufacture of genetically modified seed ensures increased dependency on transnational capital through patenting and other trade agreements (Shiva 1998), which displaces indigenous practices and knowledges. Such developments are perhaps the latest installment in the industrializing of nature and the commodification of non-human lifeworlds.

We are seeing new attempts to transform earth into productive "land," where unchartered wilderness "above" and "below" is increasingly subject to attempts to domesticate swathes of unknown territory. The concept of "terraforming" has come to be used to describe such processes where for example, celestial bodies within the discourses of futuristic space flight, become habitable through adaptation under human technology (Bryld and Lykke 1999:19–21). Less fancifully perhaps this is particularly applicable to marine environments where we have seen the development of "marine national parks," offshore drilling installations, and intensification of industrial fishing practices. We have seen in the writings of Carolyn Merchant (1980) and Vandana Shiva (1988, 1993) that the urge to domesticate land for productive purposes was a process embedded in relation of human-difference-in-domination – a gendered, classed, racialized and colonialist affair. The control of bodies and populations has shifted terrain as bio-power extends to the deepest areas of almost uncharted ocean, and beyond the boundaries of our planet. The images charted by those such as Merchant and Shiva, of a feminized and nativized "nature," also exemplify themselves in humankind's attempts to leave an unruly and unpredictable planet earth and seek refuge in a controlled terra-formation which is removed from the dangers of an autonomous "nature."

The challenge for eco-feminism encapsulated in such an extension of future modalities of bio-power is immense. The construction of cyber-earths would imply a blurring of boundaries between earth and outer space. In some scientific fields, there have been attempts to integrate understandings beyond the earth system into complexity accounts. An enormous question, then, is where the boundaries of systemic analytics might be. Whilst the politics of ecologism have encouraged us to "think global, act local," this praxis has led to a reinvention of locality and radical communitarianism. Much of this has focused on strategies of resistance to the globalizing tendencies of corporate capital, and whilst a subsistence economy may be little more than a pipe dream, there are tendencies toward substitution, reciprocity and renewable practice which tend toward a localized sustainability which is terra/forming in the best sense of the term.

Gender, nature and reproductive technologies

It is not only the natured bodies of plants and non-human animals, which have been subject to the modalities of industrial and commercial processes and increasing biotechnological intervention. Procreation, in the wealthy parts of the globe, is increasingly medicalized, and thus, controllable and predictable. Whilst some contend that new reproductive technologies (NRTs) exemplify the postmodern condition in their destabilization of scientific certainty (Franklin 1997:211), others suggest that the uncertainty of science is neither new or that the project of science as progressive control of the natural is further exemplified by experiments in conception (Oakley 2002:144). What is clear, is that there are differential effects across the globe, for as Mies and Shiva (1993) have argued the commercialization of procreation is deployed in a racist manner – designed to benefit wealthy Western (overwhelmingly white) women, whilst women in the South are often subjected to fertility control.

Some radical feminists in the 1970s were highly optimistic of the revolutionary potential of NRT's. Shulamith Firestone (1988) embraced developments in reproductive technology, infamously positing that reproductive difference was the basis for women's oppression, and women must seize control of the means of (artificial) reproduction in order to liberate themselves. Others felt rather differently, Mary O'Brien (1981) argues a naturalistic case for patriarchy as a social system grounded in the existential separation of men from species continuity, in which men's discontinuous experience leads them to seek reproductive ownership through medicalization. Andrea Dworkin argues women are controlled within a "brothel model" where they are available for non-reproductive heterosex, and a "farming model" (1983:174), which becomes more technologically efficient, enabling commodification of reproduction and a "reproductive brothel" system. Such structural accounts do not attempt to explain the unquestionable enthusiasm expressed by some wealthy Western women, for NRTs (Petchesky 1987:18). Jalna Hanmer contends that women seeking NRTs have been "blinded by science" (1987:104), and has been criticized for an inflated view of the power of medical professions (Stanworth 1987:17). Yet the medical profession does seem preoccupied with persuading its consumers that such technologies are effective, and Janice Raymond (1993:30) convincingly explains the motivation to choose in vitro fertilization (IVF) as a product of a cultural expectation of normative motherhood which fuels the desire to "try everything" in the pursuit of an elusive fertility.

The (co) constitution of procreation reflects Western and gendered notions of sexuality, and is an arena in which we are forced to acknowledge our animality in conditions of cultural circumscription (heterosexual presumption, vaginal intercourse, private enactment). Whilst the Western model of reproduction is genetic and biologized, reproduction cannot be considered outside its cultural context in kin relations and familial order (Hirsch 1999:96–118). The remaking of "nature" exemplified by current technologies of reproduction means that nature becomes increasingly artificial, but the breeding of humans has not been so straightforward for bioscience as breeding "livestock." Animals bred for meat are abstracted from their social context in the industrial complexes of most agricultural production in the West, whereas an intricate nexus of the dominations of difference frames those technologies as applied to humans.

Reproductive technologies constitute an important arena in which the cross cutting relations of "race," class, gender and nature are played out. NRTs are themselves technologies developed within animal breeding for the meat industry, including artificial insemination by donor, IVF, and embryo transfer, which is usually twinned with surrogacy (Stanworth 1987). Raymond (1993) suggests the politics of in/fertility are embedded in relations between rich and poor regions of the globe, and tell us something of the operations of transnational chemical corporations. Western infertility, traumatic though it may be, is a condition but not a disease, and it is constructed through technoscience discourse that operates to sustain a market for such technology. Historically, birth control and sterilization utilized in animal breeding, have been applied to limit reproduction of undesirable humans (poor, promiscuous, unmarried, non-white, Greer 1984:279), and this is often still the case in poorer parts of the globe. In animal breeding, the role of such technologies is to improve the qualities of the body as a potential carcass, and ensure more productive female animals are produced in disproportionate numbers.

Assisted conception does not impose any significant physical demands on men, although they fall under the criteria for patient selection and need to be relatively young, heterosexual and legally bound (Price 1999:36). IVF and embryo transfer involves complex medical procedures (Pfeiffer 1987:88) for women however. There are risks of anesthesia, surgery, trauma to ovaries and uterus, ectopic pregnancy, unknown effects of the hormones, unexpected outcomes such as multiple pregnancies and uterine infections (Price 1999:30–48) and the increased likelihood of miscarriage, congenital abnormality or caesarian delivery (Edwards 1999:44). In addition, it seems that women undergoing IVF find it highly disruptive to their paid employment and extremely stressful (Franklin 1997:82). The live birth rate per treatment cycle in the United Kingdom is 18 per cent, and death rates for babies so conceived are 32 per cent higher than those born without IVF (Oakley 2002:144). Assisted conception in humans is thus notoriously unsuccessful, indicating perhaps that treatments developed through research on one species do not necessarily transfer to another. The techniques of assisted conception were pioneered to enhance the profits of the beef industry, and across the species divide, it would seem that breeding is not so straightforward. Yet it is similarly profitable, and those with a liberal attitude to IVF should consider at whose expense ("childless" couples, stretched health authorities) technological conception operates.

In farm animals, "donors" such as bulls are kept in solitary confinement, rarely allowed contact with other cattle, and are manipulated by humans into ejaculating, or may be slaughtered for, their sperm. Embryo transfer involves impregnation with an embryo from a larger breed, leading to difficult births, and common caesarian sections (Collard 1988:116). Research sympathetic to surrogacy has found women demonstrate "grief symptoms" on giving up their babies (Zipper 1988:119), similarly to cattle (Corea 1985:237). Some suggest medical praxis is deconstructing motherhood by casting mothers in a series of functional roles, particularly as passive incubators (Hubbard 1984:350). Caesarian section is increasingly common (Mitford 1992) particularly in some parts of the United States where it can be imposed under feticide law (Faludi 1992:467). In stressing on the "natural" desire for both biological motherhood and paternal genetic continuity, reproductive technology reinscribes the biology of parenthood. Raymond suggests that parenting is being re-essentialized as it is reinstitutionalizing "male genetic destiny" (1993:35), whilst the right of women to assert the fetal embedding in their bodies increasingly diminishes. The interests served are not most usually those of fathers, who may suffer emotional and psychological damage from the impact of the repetitive failures of assisted conception practices. The interests of primary concern are those of the private corporations that provide such reproductive "services." This said we must be careful in deconstructing women's agency in the processes of assisted conception. Whilst choice is structured by a culture of reproductive fundamentalism, women can and do choose not to have treatment, to stop treatment at any time and in the limited number of cases, treatment may actually be successful. This is just Foucault's notion of bio-power as involving the internalization of the dominant paradigm, and as Catherine Riessman (1983) has suggested, middle class women have actively colluded in the medical control of their bodies regarding "infertility." This is not the case with the artifactual reproduction of non-humans, and the boundaries of anthropocentrism mean that whilst there can be no doubt that the application of animal breeding technology is problematic for humans, particularly women, it does not reduce them to "breeders." That specific formation of domination is reserved for hens and other "poultry," cattle, sheep and pigs.

For Franklin, those seeking assisted conception wish to "embody progress" (1997:96), and the IVF baby is a symbol of techno-nature in postmodern times (1997:207). Whilst the boundaries between animals and technological mechanisms are blurred as technologies are applied in cross species contexts, such practices are utterly modern. They exemplify an extreme faith in science as a modern panacea for "ills," many of which, are constructions of power. I concur with Paul Rabinow (1992) that assisted conception constitutes the apotheosis of modernity – intensification rather than transformation and contestation. It is the more "liberal" accounts, in his case, which are insensitive to social difference, and do not problematize the biologism of the motherhood-as-destiny discourse of reproductive technophiles.

A baby born as a consequence of in vitro fertilization, or a woman pregnant with an implanted embryo is a "hybrid" object, reflective perhaps of our "techno natural" times. However, to a considerable extent, human times have always been "technonatural," because as I have suggested here, we are dependent as a species, on "nature" in order to survive, and we utilize technologies of various kinds and types, in order to do so. Whatever the recent change in technologies of re/production, the iniquitous relations of difference-in-domination can be seen to make their presence felt in constituting the bodies humans and other animals.

Enfleshed discourse: the naturing of women and the feminization of nature

The discourses of domination embedded in the symbolic regimes of modernity have given rise to a range of interlocking constructions of nature, species, gender, "race," "native," such that we can say that there is a common "iconography" of Otherness (Bryld and Lykke 1999:7). We have seen in Chapter 5, how such discursive regimes are often premised upon scientific rationality which feminists, ecologists and many other social theorists have come to critique in terms of normative contents and political impact. Both ecological and other feminists have acknowledged the discourses of Othering, which draw analogies between "race," gender and nature in particular (Bartky 1979) and understand women as embodied and embedded creatures:

> Like the carbon from the air which becomes the body of the plant, and the body of the plant in her mouth becoming her own dark blood ... washing from her like the tides. (Griffin 1984:167)

In dismissing such eco-feminist interplay of discursive embodiments of gendered nature, Donna Haraway (1991), looking to a future populated with cybernetic organisms famously asserted that she would "rather be a cyborg than a goddess" (1991:181). The cyborg in Haraway functions as monstrous Other, containing narratives of domination within technoscience, but also functions as a deconstructive device for the very discourses of which Griffin speaks.

Haraway has argued that new technology and the social forms it generates problematizes and refigures the boundaries between humans, animals and machines (1991:165) so that we are becoming hybrid entities or "cyborgs" (1991:150–1). This destabilization of conceptual boundaries is a means to illuminate the false "unitary" constructions of gender, "race" and class, and suggest new ways of relating to the nonhuman (1991:170–2). However, Haraway's idealism is problematic here for whilst discourses about animals, machines and humans are culturally and historically specific, they also conceptualize "the intransitive object" (Bhaskar 1979:11), being separable from the knowledge about them. Thus cows may be seen as sacred beings, sentient mammals or food production units, but their bovine physicality remains. This is not to suggest that bovine physicality is not significantly constructed through human interventions, but that it is the same "cow" which is garlanded and left to live for thirty years, or which is fed pellets of fishmeal, slaughtered at eighteen months of age and eaten. For Katherine Hayles (1999:xiii) the relationship between "enacted and represented bodies" is a "contingent production." Technology is so intertwined with physical being that these things may no longer be separated out. Yet whatever the co-constitution of the hybrid body, I think we need the Bhaskarian position in which objects remain intransitive, in order to unpick relative degrees of corporeal hybridity.

What is particularly interesting is Haraway's slippage between modernist and postmodern analyses. She articulates a postmodern position on the boundary blurring between animals and machines, almost suggesting the collapse of conceptual distinction. "OncoMouse," the world's first patented mammal (genetically engineered by chemical multinational Du Pont) is a research tool born with cancer bearing genes. Haraway describes such a creature as a "technobastard" (1997:78), a "primal cyborg figure for the dramas of technoscience" (1997:52), a biotechnological "actor" that "cooperate(s)" with cancer molecules (1997:97). She sees the combining of animal and plant genes as evidence of our promisingly cyborg future in which mice, rice and tomatoes become "fully artifactual" (1997:108). Haraway ignores any exploitative power humans exert on the basis of species membership, and denies any autonomy or agency to the environment. She interprets environmentalist opposition to genetic engineering as "racist" in its concern with "purity of type" (1997:59–60), and describes the anthropomorphizing of chimpanzees, mice and rabbits in bio-technological product advertising, as "wonderful" due to the cyborg imagery of blurred categorization (1997:254). There are no questions raised as to the material abuse of sentient mammals in experimentation (see Birke 1994), or of the gendered, natured and racialized content of their discursive representation. However, advertisements that carry gendered and racial discourses in relation to animalized humans are "ominous" (1997:255–65), and she retreats into a modernist analysis of the exclusionary interces of gender and "race" when discussing the symbolic regimes of the human. She is anthropocentrically unable to see social domination when looking at natured relations, and this is reflected in her postmodern account of the application of biotechnology to non-humans.

In contradistinction, the commodification of non-humans, is key to Shiva's analysis of biotechnology wherein plants and animals become: "instruments for commodity production and profit maximization" (1998:29). Globalization involves the "predation" of one class, "race," gender and species on others wherein the "dominant local" seeks global control (1998:105,122). Organisms and eco-systems lose bio-diversity, domestic animals suffer abuses in genetically engineered meat production, bio-technological seeds result in superweeds and superpests (1998:37–41), and third world agricultural communities are disempowered by the racist imposition of patenting restrictions and forced to import Western seed (1998:59). "Modernist" conceptions of class, "race," gender and "nature" are implicit in Shiva's analysis, which examines both material changes in technological applications and their effects, alongside the shifting discourses about "nature" and "society" with which they are associated. Contemporary forms of hybridization, whilst involving physical interpenetration across species, do not easily contest power, but often remain embedded in the social networks of domination based on species, gender and racialized difference.

There are persuasive feminist accounts of the discursive co-constitution of gender and nature that suggest the continuing co-constitution of domination, rather than its contestation. Carol Adams's work on gender and species relations focuses on the interpellation of symbolic discourses of gender, nature and "race" with regard to the bodies of women and animals as commodities for male consumption. Adams has integrated analysis of material practice and context with symbolic regimes of hierarchical difference, arguing that in battery, domestic rape and femicide (Radford 1994), violence against domestic "pet" animals is a strategic expression of male power and a means of control over women and children (1995:76–8). However, she is best known for her observations about contemporary meat eating symbolic cultures. For example, animals are "absent referents" in the mass term of meat, she suggests and the gendering of this process is threefold (Adams 1990:26). First, women may be represented through the lens of butchery, as fragmented body parts which are sexualized for pleasurable consumption, or through the animalized female referent (bitch, pig, bird). Second, species, as it applies to agricultural animals, is gendered, for what we eat as meat are "females, castrated males and babies" (2003:132). Most recently, she has made a powerful case that the cultural spaces and places of male fraternity are arenas policed by homophobia and a particular attitude to the dominations of nature and gender. Adams's analysis is based on the deconstruction of representations of food and its consumption and there is scant reference to the material domination's constitutive of the processes of farming, slaughter and butchery by which animals become meat, but, like Shiva she draws our attention to the complex structures of gendered and racialized domination that is implicated in the species relations of contemporary globalizing capital.

In this context, where does Haraway's mouse sit? OncoMouse™ is owned, incarcerated and programed with disease. It is also a theoretical waif. Haraways cyborg connotes a plethora of creatures having "no truck" with "seductions to organic wholeness" (Haraway 1991:302). Yet this cyborg "motif" is a distinctly disembodied and topian figure abstracted from its corporate biotechnological context. Postmodern understandings of bodies, be they human or other animal, is considerably naive in failing to recognize bodies as mapped by cross cutting lines of gender and other formations of power, and indeed, as Foucault suggests, as modified, inscribed, proscribed, infringed by such complex relations.

Contesting hu(man)ity: toward an enfleshed posthumanism

This book has no real space to consider "what is to be done" – that is another project. Given the complexities of the intricate webs of dominatory relations, solutions are not easily found. There have been many flights of imagination and fancy, however, wherein we may try to comprehend a little more of the complexity of being Other, and of being *another* "Other." We don't require fiction for such "world traveling" for Others undertake such practice as necessary daily experiences (Strathern 1999:189). As women, as minority ethnic, as relatively poor, as less able bodied, as homeless, landless, stateless, in opposing warfare, abuse of other animals, peoples, spaces and places, we are enmeshed in contradictory and complex relations to dominatory power formations. The wider we theorize complexity, the more complicated and difficult our political decisions become. Yet we must take on board our problematic insider/outsider political status and struggle for inclusivity, often against our own exclusive desire. This final section explores a few symbolic and material enactments of the contestation of dominatory power in order to suggest that whilst the matrix of complex domination is dynamic, robust, flexible, multifaceted and multilayered, there are projections toward a world in which the complexity of all our difference may not be so difficult and often, dreadful.

Postmodern influenced feminism and ecologism has emphasized the importance of blurring the thresholds and boundaries of exclusivity around difference. Haraway, as we have seen, attempts to deconstruct biology in asserting that "nature" is a social construct. She is provocative in questioning feminist resistance to seeing women as animals, and like Plumwood (1991:19) sees this as a deployment of Enlightenment categories. Drawing on Deleuze, Braidotti (2002:118) deploys the notion of the "becoming" self in order to understand the connection between the "becoming-Others" of modernity. What Deleuze (1979) seeks in his notion of "becoming-animal" is a tool with which to engage our human animality. He problematizes the Western Enlightenment notion of the (rational) embodied subject as inhabiting a perfectly functional physical body, and suggests that bodies that are pathologized (disabled, aging) are stronger clues to the "monstrosity" of our animality. Animals are mechanisms for change, for metamorphoses (Deleuze and Guttari 1975), yet for all the emphasis on the corporeality of animality, real animals, the corporeality of their Otherness in modernity, as machines for milk, fat, meat, transport and protection, are absent. The becoming-animal is a metaphor, whose body is analytically superfluous, thus some feminists have anthropomorphized Deleuze's preoccupation, insects, as "queer subjects" (Grosz 1995) without seeming incongruity. My own approach to understanding Other animal natures is very mindful of the ways in which discourses of natured Otherness are exemplified in the practices of capitalism, colonialism and patriarchal conditions of life, and the way such social forces bend, transfigure and often break, the bodies of non-human animals. Braidotti provides an idealist, cultural analytics of the use of animal-metaphor to understand human, particularly female, experience (2002:127-31). Her understandings of animality reproduce the very human/animal binary she claims to wish to deconstruct.

Following Deleuze, she considers a range of animal practices, behaviors and habits as if these are not also human habituations (2002:133). In addition, she conflates "animal" without considering the very humanist anthropocentrism, which frames the category in the first place.

Some feminisms and ecologisms have been engaged in the problematization of this category "human," and this is of course, a thread that has run the course of this book. Judith Halberstram and Ira Livingston (1995:3) suggest that, given the problematics of biological explanation and the social constitution of both humanity and animality, the embodied human is best seen as part of a "zoo of posthumanities." Both Haraway and Deleuze question the continuities and discontinuities between the human and non-human, they are concerned with the embodiment of Othering, but fail to unpack their technophilia as a discourse located firmly within high modernity's quest for mastery over earth and space and all the other Others of modernity. Neither do they consider the deep-seated anthropomorphism embedded in their analysis of animate and inanimate "machines" and other less strongly modified cyborgs. The cyborg is defined by its human manipulation or the incorporation of technology within the human body to serve human ends. Yet the running of human–animal machine in linear fashion and as conceptually interchangeable is problematic. The projection of human and animal bodies into an "artificial environment" is a very old story of cosmic redemption told with the accent of postmodern disembodiment. We do not inhabit an entirely technological habitat, and to presume so is a product of a very partial perspective (see Shaviro 1995). Human cyborgs are often undesirable – they are the product of disease, illness, near fatal injury. The social condition of becoming cyborg cannot be separated from the operations and machinations of modernity as a project of complex difference-in-domination structured in complicated ways by relations of capitalism, colonialism, patriarchy and anthroparchy.

Technophilia and technophobia is another modernist dualism. Whilst some ecologism and radical and ecological feminist theory demonstrates elements of a nostalgic imaginary, there are strongly utopian elements of Haraway's (1991) best-known work where she makes the case for a shift from industrial modernity in arguing that "white capitalist patriarchy" has been replaced by "an informatics of domination" (1991:162). Capital in contemporary times is arguably more powerful than ever (Callinicos 1994, Hirst and Thompson 1996, Sklair 2000). An "informatics of domination" is also one in which racism may be intensified, given the continued dominant of standards of "whiteness" (hooks 1992). Any posthuman notion of the human body, must take account of the problems incurrent in shaping our bodies through modernity in terms of class, race, sexuality and gender. Perhaps most important is the need to question the liberal individualist notion of the body as a repository for the self, the autonomous subject whose "manifest destiny is to dominate and control nature" (Hayles 1999:288). This said, notions of the posthuman may open up the possibilities of new kinds of dwelling in what Naess would call "mixed communities" of animals and technological formations/hybrids. Perhaps of most potential in refiguring relations of domination in a posthuman future, are the roles of other animal species, hybrid, for the most part, but able to operate beyond human control in a manner far more unpredictable and challenging of anthropocentrism than the quirks of any robot or computer.

From cyborgs to companion animals: different trajectories of the posthuman

In the "cyber" literature, real bodies become "meat" from which virtual reality enables one to escape (Sobchack 1995, Lupton 1996). For Hayles, the posthuman can be defined as a privileging of "informational pattern over material instantiation" (1999:2) which can be both disembodied and virtually embodied, in presenting the human body as "the original prosthesis" which we learn to manipulate. The posthuman worldview sees no essential differences between human bodily existence and computer simulation, between organisms and cybernetic mechanisms, as humans and other animals are seen primarily as information processing machines. This is not necessarily a "new" development, for just as Latour suggests that we have never been "modern," so too have we never been "human," or not in an embodied sense. Feminist and postcolonialist critiques of universality have countered the predilection in Western social and political thought for the autonomous white man of liberal individualist humanism. This liberal humanist subject was also disembodied – he possessed a body, but was not an embodied creature.

Haraway assumes that the posthuman cyborg subjects will experience their posthumanism, and the utopianism in her account is reflected in the notion that this is enriching and enabling in some way. Yet the notion that "we" become posthuman because we think we are, is deeply unsatisfactory. The posthuman in this sense of seamless interplay between human experience and cyberspace, is a highly specific and located phenomenon. The notion of the posthuman as a "condition" is a universalistic presumption, for any subjectivity of the posthuman affects an incredibly small proportion of the world's population.

As Hayles points out, both human and posthuman are historically specific constructions emergent from different formations of embodiment, technology and social organization (1999:34). Haraway would do well to attend to the situatedness of her posthuman utopianism, for the retreat into cyberspace is from the body prone to disease, to decay, and embedded in ecological spaces and places that may be increasingly uncomfortable.

Whether the notion of hybridity is both flexible and specific enough to capture the qualities of social nature is moot. For Haraway, the biological conceptions of organism and mechanism are rendered obsolete, and for "post-industrial people," organisms are constructions increasingly of an artificial kind (Haraway 1991:208). Yet as Birke points out, organisms do not have the fluidity Haraway suggests. They are entities characterized by processes of self-organization and a relative stability of form over time. In addition, "we" are not all post-industrial, and whilst certain communities within first world societies may be enmeshed within information networks, the hybridity of the poor assumes different formations, and these may be distinctly embodied. There has been debate as to the extent to which the cyborg may be seen as a liberatory figure for feminists (Stabile 1994b, Lykke 1996), and critiques are pertinent to her understanding of the body. Haraway loses the materiality of human and other animal corporal being, for organisms become codes of information which are constantly constructed and reconstructed, "strategic assemblages" which are "ontologically contingent" (1991:220). Such contingency has corporeal specificity however, depending on the location of bodies in geographies and socialities of difference. Medical cyborgs exist, with pacemakers, and pig valves in their hearts. The possibilities for both these kinds of limited cyborgs and the more usual connotations of computer-aided technoculture are evidenced in different forms and degrees across the globe. The postmodern and posthuman flight from the flesh and the system boundaries of an organism has its own essentialism and utopianism, reflected in fantasies of disembodiment and the "overthrow of matter."

Haraway has always given credence however, to some level of ontological realism in allowing "nature" and its multiplicities of species an agency as "material-semiotic actor" (Haraway 1991:200). Before she concentrated on the "cyborg" as a contestationary symbol of the hybridity of social/nature, Haraway tended to characterize nature and its independent trajectories as a "witty agent" compared to the North American prairie wolf that plays the role of trickster in Native American mythologies. Haraway sees the coyote as hybrid, living on the borders between human society and the "wild," and as such, non "essential," albeit the wild is perhaps best seen as a human interpretation with differential degrees of hybridization. Haraway has once again, gone to the dogs, arguing that domestic dwelling dogs or companion animals "more fruitfully inform livable politics and ontologies in current lifeworlds" (2003:4) than cyborgs or coyotes. Bryld and Lykke (1999:183) have posited the dolphin as an iconic figure for contesting mulitiplicitious dominations including environmental destruction and militarism. Dolphins are presented as nomadic jesters with which we might explore the species boundary. For Haraway such animals are not "an alibi for other themes," not "surrogates for theory" nor indeed, for other relationships in a "multispecies family" (2003:96), instead they are "fleshly material-semiotic presences" which are "here to live with" (2003:5) rather than just to think with.

I consider Haraway to be moving into an embodied interspecies form of posthumanism here. Her *Companion Species Manifesto* provides a nuanced account of the historical relationship between humans and dogs. This relationship "is not especially nice; it is full of waste, cruelty, indifference, ignorance and loss, as well as ... joy, invention, labor, intelligence and play" (2003:12), but it is certainly also an account of what Lovelock would call "co-evolution" (2003:26–32), and she references symbiogenesis as a rule of genealogical thumb. The material-semiotic presence of dogs can be evidenced in their use as labor for herding, guarding and hunting, and the role of "pets" as household members and even bed companions. Discourses of racial purity infuse socialnatures in the breeding of dogs and the narratives of their "civilization," alongside the practices of welfarist bio-power. Problematically, in considering the Otherness exclusively of companion species, only a partial story of hybridity is told here, and the status insecurities of companion animals in a context of human species domination are left untheorized. Hybrid relations of dependence, affection and fun, may be icons for a posthuman future, but the structural basis of our domestication of animals needs accounting for. Haraway's intention is to draw together the multiplicities of difference, often marked with the notion of "significant otherness" (2003:7). She plays with the deconstruction of "species" but ultimately hangs onto the need for taxonomy of species difference (2003:15), whist being specific in defining "companion animals" as those in relations of "biosociality" with humans as workforce or family. Although she castigates "deep ecologists" for some supposed equivalence of treatment of animal others, this is rarely found, and most would accept an "honour" of difference (2003:39), and of communication "across irreducible difference" (2003:49).

Both human and non-human animals possess autopoesis and cognition and organize their "living" in relations of reciprocity and interdependence (structural coupling) with their environments. Complexity eschews an analytics of embodiment based on genetic code, and emphasizes that the nesting of systems within systems joins networks of variate species in worlds of cognitive awareness. In our complex multispecies communities, the liberal humanist subject is not stretched, nor fluid or blurred at the boundaries, but is revealed as a myth (Valera *et al.* 1991:98–9). The posthuman, ultimately, may not be as "post" as it may first appear. Whilst there are comparisons to be made between the conditions under which both computation and life itself are likely to emerge, this does not mean artificial and human/animal life are the same. Whatever the fantasies of escape from the flesh implied in the narratives of cybernetics and artificial life, as Varela (1991) notes, the human mind is embodied, and without the body we are not human. To suggest we might be reduced to the informational content of the brain alone, is to uncritically co-opt individualist humanism into the posthuman. Cognitive beings are alike in their embodied realization and actualization of self(hood). Haraway desires a means of relating beyond the human which is cognizant of the multiple and complex dominations which shape formations of society–nature relations, or naturecultures. In the process of that realization, the dark side of natured domination needs more exposure than she is prepared to give it. Braidotti (2002:261) ends her account of the body by advocating what she calls "strategic essentialism." Both feminism and ecologism have deployed bodies in effective political ways in resisting the powers of gendered and natured capital. But we do not need real bodies just for political resistance, but for critically engaged theory too.

Theories reflect social and cultural values and are generated from social networks and within systems of complex domination. Complexity theory cannot liberate us from this condition, but it offers us a better range of concepts with which to understand our embodied condition, the structuring of the body and the embodiment of structure. There is a difficulty with the matrix of complexity, in that systems within systems of interlinking and interlocking biosocial difference can be understood as precluding change. Yet systems are never hermeneutically sealed, and in their dynamic process change might be envisaged. Leaving "the meat behind" is impossible, for to be human is to be embodied. The boundary to the seamless integration of human with the machine is the distinct difference of their embodiments. Ecological feminisms have questioned the imperialist project of the domination of nature, and the fantasies of gendered disembodiment that accompany such processes of subjection. Yet the posthuman may not necessarily be anti-human or anti-animal or disembodied. It may pave a way for different understandings of relations between species and all forms of "life," and of the processes of life itself. Respect for the multiplicities of difference becomes ever more complex as the diversities of both dominations and possibilities emerge. Biotechnology has opened a path of possibility that in questioning human centeredness, may lead to an embodied and ecologically embedded posthuman future, but we cannot simply dissolve the categorization of human and non-human which is an edifice of political power, economic organization and social stratification and distinction. Rather, we need to examine the complex interrelations of difference-in-domination, in order that we might find the chinks in the networks and nodes within which this might be realized.

7
Domination in a Lifeworld of Complexity

> ... my dream is a version of the posthuman that embraces the possibilities of information technologies without being seduced by fantasies of unlimited power and disembodied immortality, that recognizes and celebrates finitude as a condition of human being, and that understands human life is embedded in a material world of great complexity, one on which we depend for our continued survival.
>
> N. Katherine Hayles (1999:5)

We are at a point where the undermining of the metanarratives of contemporary social theory and the erosion of disciplinary boundaries facilitates, Niklas Luhmann suggests, the development of a new theoretical paradigm. That paradigm is complexity. This is beyond *a* theory, and has been described as "supertheory," yet a complexity understanding of the lifeworld does not imply some kind of right to "truth" but is a position of contingent truth, which is inclusive of the "reality" it describes. Complexity theory is "big" theory but whilst contemporary physics stretches itself toward a theory of "everything," which would explain the entire physical universe, including its own possibility and existence, a theory of social domination seems rather modest in comparison. Some will consider what follows to be "grand," not least in its attempt to capture the intermingling of nature, gender and other mutiplicitous differences-in-domination.

This concluding chapter draws together my conceptual threads and theoretical arguments. I begin by discussing the different levels of abstraction involved in articulating a complex multiple systems approach of difference-in-domination, and considering how these differential abstractions relate to each other. Much of this book has been devoted to a defense and deployment of concepts such as system, structure and also discourse, and pulling through these concepts within a range of literatures from ecologism, feminism and ecofeminism. I drew on elements of all the different varieties of ecologism, and on complexity theory in the natural sciences, in arguing for a social system of domination of nature, anthroparchy. I defended structural and system approaches to gender relations, whilst arguing that a system of patriarchy should be conceptualized in the context of other systems of social domination. I have mapped the range of ecofeminist understandings of the relations between gender, nature and other formations of domination, and suggested these might be drawn into discursive, structural and systemic analytics, whilst also accounting for our embedding in non-social contexts and the complex structuring of human embodiment. The second purpose of this chapter is to suggest the elements of systems of patriarchy and anthroparchy in order to specify the range and limits of each. In order to do so, I outline various social structures (groups of oppressive relations) of which such systems may be composed. Some of these structures may be common to both systems, despite differing content, whilst others may not be. Third, I consider which system elements might interpenetrate, and which remain relatively exclusive to one particular system, in order to exemplify more explicitly my arguments for a matrix of complex systems. Finally, I consider the possibilities of change, both drastic and incremental, in unpicking the threads of the matrix of the multiplicities of social domination.

Complex lifeworld

Our planet is characterized by enormous difference. Its biota and scapes are so plentiful and diverse as to be unknowable. Even in terms of human society, our social formations and practices are incredibly varied, shaped by the specificity of the histories and geographies we inhabit. Yet such diversity does not preclude patterns and forms, which whilst dynamic, exhibit some degree of continuity and stability over time and often also, space. The arguments presented in this book draw on complexity theory in both the social and "natural" sciences. In both cases, social and natural lifeworlds are conceptualized as operating with sets of relationships between phenomena that can be described as "systems." I have focused on social systems of power relations that are constitutitive of relations of "gender" and "nature," but a whole range of systems might be seen, operating locally or more widely, and as being constitutive of various kinds of social domination based on hierarchies of difference.

There are other systems to which I have referred, such as capitalism, and postcolonialism that significantly influence systems of gendered and natured domination.

Social systems may be reflexive and exhibit self-organization and adaptation. Both these qualities are related to the abilities of social systems to be dynamic and interrelated or "interpenetrated." Luhmann refers to such dynamic interrelation as *"hypercomplexity."* I have argued that patriarchy and anthroparchy should be conceptualized as relatively autonomous systems of social domination, which may articulate and "interpenetrate" and as such, that a multiple systems approach might be efficacious in analyzing such interrelations. There may well be more than four social systems of such significance that they shape social relations from the local and specific to the global and/or planetary level, and some may disagree with my specification of the systems with which I deal in depth. My main intention however, is to advocate complex systems analysis as a tool with which we might capture the diversity of formations of social domination, and which enables us to understand these forms as embedded in and also co-constitutive of, some non-social systems.

Discourse, structure and system

Three stages of abstraction have percolated through the pages of this book, and they constitute different but interrelated stages of theory building. At the least abstract level, *discourses* are generalizations about symbolic representations and their articulation through material institutions and procedures. These generalizations can be grouped as discourses, through their thematic interrelations, and conceptualized as "ideas-in-practice," which carry and constitute power relations. I see discourses as sets of ideas that are institutionally embedded in social practices, and which are both constructive of the social world and constitutive of oppressive relations of power – attaching patriarchal and anthroparchal ideas to social, economic and political institutions and practices. I argued that discourses should not be seen as heuristic devices, but as applied practices with real effects. Discourse analysis is both less and also differentially abstracted in relation to the structural analysis I draw upon. I consider that this is Foucault's own sense – the production of a local, specific, detailed and fragmentary knowledge that is able to catch the complexities of detail in operations of power. I also suggested that such analysis might be combined with "macro" theoretical explanatory frameworks of structure and system.

Discourses might be seen as operating both within and across groups or sets of relatively enduring oppressive relations of power – *structures*.

Structures can be evidenced in more concrete instances than the further abstracted notion of system, and are constituted from sets of social forms such as institutions, and from regularized behaviors and practices often relating to such institutions. Structures are deep-seated sets of institutional/organizational/procedural relations, which shape social life in important ways but do not determine it. Whilst societies are composed of structures of social relations, human beings have agency which impacts on such structures, for structures both survive and reproduce themselves, and are also contested and altered, by human action. The significance of agency may differ quite profoundly across different oppressive systems. In the domination of certain animal species, for example, almost all humans are implicated as agents of domination. The majority is unreflexive as to their role as agents of domination, but there are instances in which humans can be anthroparchally contesting agents, and the extent and degree to which humans exercise domination is specified by location in place and time. Systems and their constitutive structures of oppression are not static, changing in form, degree and mode of operation over time and across cultures, and oppressive structures are both reproduced and changed by human agency. However, it is difficult to envisage a social theory of anthroparchy that is able to account for agency in nature itself. An understanding of agents' thinking enables an analysis of agency, and understanding the cognition, never mind the "perception" of non-human biota is problematic. Complexity theory in the natural sciences does enable a view of cognition in all life forms which is self-making and replicating and in the replication we may have adjustments of form. The social domination of nature inevitably suggests a system embedded, co-constitutive of and affected by, physical, chemical and biological systems, but my understanding of "anthroparchy" is specifically of a *social* system of institutions, practices and social formations that organize human interaction with "nature." As such, the agency of which we might speak when thinking of designed change (rather than that which is emergent) is that of human beings contesting and reshaping anthroparchal relations. The natural world is very much characterized by entities and by dynamic systems, which have their own emergent properties and powers and these are sometimes (in the case of global warming, for example) strongly implicated in social systems. However, such interrelations are best seen in terms of the interpenetration of social and natural systems, rather than an anthropomorphic projection of agential nature "biting back."

I conceptualize structures as more abstracted from empirical instances than discourses, which are emergent from/within such instances.

Discourses and structures are also differential abstractions, the former being more detailed, specific and fragmentary. Whilst discourses operate within structures and construct and constitute structural relations of power, I do not see them as discreetly contained within structures, but as evident in differing form and content and operating in complex webs of interrelation across structures. This may give the impression that I am conceptualizing structures as rigid boxes across which discourses may flow, but this is not the case. I do see structures as relatively enduring sets of relations, but whilst they are distinct, they are not autonomous, and structures link, overlap and intermesh in certain instances. Such links, overlaps and intermeshing, and a sense of distinction combined with semi-autonomy, characterize all my stages of conceptual abstraction.

Structures, in their interrelations, form complex systems of oppression. Such systems can be seen to subordinate certain populations in webs of oppressive relations. These systems do not exist in isolation, and contemporary "Western" societies are characterized by a number of systems. At some levels, these systems can be conceptualized as operating in similar ways. Some patterns of relations of capitalism, postcolonialism and patriarchy can be seen to operate at a global level, such as production and exchange, the movement of goods and labor. The specificity of power relations is often best conceived more regionally, and we might speak of varieties or kinds of capitalism and patriarchy for example. The specific social formations, the structures of institutions and practices that organize our relations to "nature" differ substantially across the globe, and nature might be dominated to very differing degrees. So, I do not think that these systems should be conceptualized as parallels. Systems of oppression are likely to have particular structures that are specific to them. They may have structures that can be seen as similar or even the "same," but the content of those structures is likely to differ in terms of their specific formation and in the degree of oppressive relations that constitute them. For example, violence can be seen as a structure of both patriarchy and anthroparchy but the form (type) and degree (level) of violence differs. Thus mass killing is not an endemic kind of violence for women in Western patriarchy, but is an endemic form for certain groups of domesticated animals in Western anthroparchal society. Structures of violence operate to differing degrees: for example, anthroparchal violence is more intense than patriarchal violence due to rountinized slaughter of some domestic animals for food. Systems of domination relate in ways both co-operative and conflictual, and disparities and similarities between systems of oppression might be explained via an examination of their structures.

Patriarchy, anthroparchy, capitalism, postcolonialism and other existent and emergent systems of social domination, exist within a *matrix of hypercomplexity*. Whilst systems of domination should not be seen as parallel phenomena, I consider that they are likely to exhibit key elements common to complex systems. First, *systems of social domination* are characterized by "autopoiesis." Such systems are self-making and have networks of components which produce/reproduce and transform other components, leading to the maintenance of the system as an entity. *Autopoetic organization* includes the creation and maintenance of system boundaries and the constant renegotiation and creation of relationships within the system. However, social systems should also be seen as exhibiting both *open and closed properties*. Social systems can be open to new material, whilst at the same time maintaining their form. This means that systems have autonomy, and are self-organizing in the sense that their form and pattern (or structures and operation/behavior) is not imposed externally, but generated internally. This does not mean social systems do not interact with other social systems and with the systems of the physical world, but they are open to a constant flow of material. This relation between pattern and change explains why social systems, like natural systems, engage in the process of reproducing their forms and relations, and in doing so, shift and re/form. As such, social systems, like natural systems are in a condition of coexistence between change and stability, the chaos of change and the order of structure. Social systems then, can be said to *restructure* themselves continuously and are dynamic.

In the natural world, systems are characterized by *emergence*. This is not teleological, for it is only great apes, that is humans, chimpanzees, bonobos, gorillas and orangutans, which possess species-specific properties of design. Most natural structures are not designed, but emerge from the processes of co-evolution and are novel, creative and flexible. Human designed structures, being "planned" in a loose sense, are capable of dynamism and change, but tend to assert control over both human and other lifeworlds. Whatever the assertions toward the imposition of order however, a complexity understanding of both systems and structures means that we can speak of social systems and their constitutive structures emerging at higher levels of complexity as they reflexively incorporate additional material, changing contexts and undergoing internal shifts. The emergence of increasing complexity is a particularly prevalent characteristic of human societies however, because unlike ecosystems, social systems have components (people!) with a relatively high degree of autonomy, who are crucial in replicating and changing those systems.

In addition, systems of social domination should be seen as *nested* within other systems. For example, patriarchal systems of social relations of domination are nested more widely within "society." Society is multivariate across the planet however, and is implicated in equally varied natural systems or eco-systems of other species and life forms. The specific contexts then, of nested systems, influence (but do not determine) the constitutive structures and their articulation in specific region and locale and we can say that social systems are *embedded* in an environment of multiple and complex systems. Systems of social domination may be seen as nested in broader social systems and in relations of co-evolution with regionally and historically specific natural environments. In addition, although a defining characteristic of any system is the maintenance of a boundary, systems are *interpenetrated* by other social systems, which collectively form their context. In examining any particular form of social domination, we must expect the intersection of various other forms of social domination that increase the internal complexity of a system. Systems of social *domination* exert controls in very different ways through inclusion and exclusion, marginalization, or through harsher and more obvious exploitation of labor or life as a resource, and through the oppression of human and certain other species that are prevented from flourishing in species-specific ways.

Thus discourses, structures and systems constitute different stages of theory building, each stage being progressively further abstracted, whilst also being differentially abstracted. Whilst I conceptualize systems of oppression to be composed of and constituted by structures, and structures in turn to be composed of and constituted by discourses, I do not see the abstraction of oppressions in terms of neat and discrete categories. Rather, at each level of abstraction, the concepts are seen as interlinked and overlapping in ways that I think is appropriately reflective of the complexity of social life.

Multiple systems – a matrix of social domination

I have argued for the adoption of a *multiple systems* perspective. Specific instances of social domination are complex, and likely to be produced by the articulation and interrelation of discourses and structures constitutive of more than one system. In each instance or case, a varying number of systems may be present, operating to different degrees. For example, socialist feminists have seen domestic labor as part of the structure of the household in capitalism and patriarchy. It is affected by "race," although not necessarily in negative ways – for example some black feminist theorists have defended black families as less reflective of gendered disadvantage, and in contestation with an ethnically structured society. I think it is most unlikely that the household can be seen as either a structure of, or a phenomenon affected by anthroparchy, other than by its location in the nexus of consumption. It is not imperative, nor often possible, in explaining particular instances of social domination, to refer to all potential systems. Complexity theory enables us to conceptualize social systems that are non-teleological, continually dynamic, interactive and fractured. I have compared the complexities of social domination to a *crystalline* formation, adopting particular forms, alignments of planes in specific formations, which can apply to particular instances or empirical cases.

I have also described the intermeshing of different systems and structures of domination as a matrix. *Matrix* captures this sense of the networks, of intersections between phenomena, in this case both bounded and also permeable systems of social domination and the structures of which they are constituted. Drawing on a number of meanings of a "matrix" we can conceptualize the field of social domination as contained within, or as constituting, the entity of matrix in the following ways. First, the extent to which different systems of domination cross cut and intermesh means that they coalesce in a particular array or pattern. Interrelation is both a strength and a vulnerability within the field of social domination, but the cross cutting of systems also means there is no one, specific "way" out, although there are myriad chinks, weak points and arenas for negation and (re) negotiation of power. Thus, second, we can draw on the meaning of matrix, as per its Latin derivative as both "womb" and "pregnant animal." In this sense, the matrix is not a fixed entity, but an arena of possibility, and becoming, of development and process. Social relations are not simply cast in a way that reflects the patterns of systemic properties within the field of domination, but are continually recast. Systems of domination restructure themselves continuously, and such structuring incorporates positive and effective challenges to systemic power that alter the ways in which structures of relations operate, and the degree to which they can be said to be exploitative, oppressive, exclusionary or marginalizing. Third, a matrix is a descriptor for both surroundings or environment and the context within which something is formed. We can draw on this notion in understanding the layered quality of system relations within and without the social field. Complexity theory suggests that systems are nested, existent one within another, and that both social and natural systems exhibit such properties. This notion can be applied to capture the embedding of social domination within the contexts of physical systems, including the "natural" and not-so-natural, environment. Social systems are embedded in such contexts, and exist in dynamic interaction with non-social systems, both influencing them and being constituted by them in turn.

Realism, embodied materialism and symbolic regimes

Systems of patriarchy and anthroparchy and their constitutive social structures and discourses of power have a real existence and effect. Systems, structures and discourses are not only heuristic devices a social theorist may use as a tool to explain phenomena. Rather, they are properties that can be ontologically established via empirical investigation, for they have emergent features and powers and corporeal effects. These emergent properties can be seen to "exist" regardless of our interpretations of them.

I argued for example, that the environment refers to specific physiological entities that should be analyzed in terms of specific systematic structures that are real and have real effects. I also contended that my adoption of critical realism was likely to necessitate a structural approach to the analysis of a social system of human domination of the environment, for the multifarious natural environment in an anthroparchal society was dominated and controlled by humans. I linked my arguments for a realist ontology to those that held a structural approach significant for the analysis of gender relations. Here, I contended that whilst gender relations should be seen as dynamic, they exhibit regularity and continuity over time, and have a real existence beyond our knowledge of and enactment of them. Thus the discourses, structures and systems that I have discussed, I consider to be real objects with emergent properties that may help us identify why a particular phenomena may be, as Andrew Collier puts it, "thus, and not so."

Real dominations that are exemplified by social systems operate at a number of levels. The specific, detailed elements of gender and nature as complex regimes of ideas coalesce in what I call symbolic regimes of normative belief that is at once both separable from corporeality and closely intertwined with it, to the extent of being constitutive. Such symbolic regimes are both reflective and constitutive of oppressive power relations, they are embedded in social institutions, practices and processes and shaped by various systems of domination that construct and legitimate dominant power relations.

Structures of systems of social domination are conceptualized as having ideological and material aspects or levels, which can be differentiated and are also interrelated. The material level is where discourses of oppression are concretized in physical form. Whilst this can refer to economic institutions and processes, these can also be political or violent. For example, the state can be seen to operate both ideologically and materially. Legislation deploys cultural discourses of gender and nature, but law is materially present in institutional form, practically enacted in gendered and natured ways by police, courts and punitive measures. Patriarchy and anthroparchy assume different forms and degrees in different levels of analysis. Systems of domination may be seen to interconnect at different levels, and may appear in some instances and not in others.

Ecological feminists have been very much concerned about the environmental impacts on women and children's bodies, and on the gendered particularities of the embodied experiences of women. I have argued for a complexity understanding of our embodied human condition as part of a materialist analytics. Bodies are dynamic and self-organizing systems, in a constant process of transition and becoming, despite the stability of their external form. In turn, our embodied human condition means that human bodies are dynamically engaged with the physical and social systems, which constitute the "environment" of the individual and collective body. Social systems impact physically and symbolically on bodies both human and non-human. Our sybiogenic relation to the "natural" world of animate life forms and inanimate contexts means that we are implicated in environments by acting on/with them in order to survive. We modify and technologize other bodies, spaces and places, which means much of what has been thought of as "nature" constitutes a multiplicity of hybrid entities and formations. Just as "we" humanize nature, so are we constantly subjected to processes of naturalization, as human animals in a condition, as an animate species, of co-dependency with our environment. Recent developments in hybridization have led to a debate on the instability of boundaries between humans, animals and machines, but any talk of a posthuman condition is problematic unless it foregrounds our embodied and embedded condition of material being.

Real dominations in social systems

I have argued for a structural approach to the analysis of gender relations that operationalizes a conception of a system of *patriarchy*. However, in explaining gender relations, the interpenetration of multiple systems means that specific forms, practices and collective experiences will demonstrate multifaceted complexities of differential dominations. Gender relations may most usefully be seen to articulate in institutions, processes and procedures that can be conceptualized as structures. Such structures have certain effects and can be considered real, and may be both trans and counter phenomenal. Yet people have choices and options, they may act as agents of reproduction within patriarchal structures or may also contest and change them, for agents are implicit in normative praxis, and restructure social institutions, practices and relationships, and thus are systems autopoietic and dynamic. Patriarchy is a system of social relations based on gender oppression, in which primarily women, but also feminized Others (children, insufficiently patriarchal men) are dominated and oppressed. Structures of patriarchy are linked, but have relative autonomy. Patriarchy is characterized by different structures that emerge from normative praxis. These structures are based upon aspects of systems of domination. I concurred with Walby's (1990) identification of six structures of patriarchy: paid employment, household, culture, violence, sexuality, and state, whilst disputing her prioritization of certain structures, and her marginalization of both the embodiment of structure, and of the ecologically embedded constitution of the social. I suggested the structures that interpenetrate with anthroparchy are: sexuality, culture, violence and the state.

The symbolic separation of human "culture" from "nature" is constituted by and through social institutions, processes and practices – through social structures. These are sets of relations of power and domination that are consequential of normative practice and interrelate to form a network, a social system of natured domination or *anthroparchy*. Anthroparchy is a complex system of relationships in which the "environment" is dominated by human beings as a species, and it involves different degrees of the form and practice of power: oppression, exploitation and marginalization. I use these terms to capture the different degrees and levels at which social domination operates, and also the different formations it assumes, within which some species and spaces may be implicated whilst others, inevitably, given their degree of difference, are not. Whilst deep ecologists have been right to argue a case for "anthropocentrism," this "centrism" can capture forms of human–environment relations that exemplify marginalization, but is rather weak for the capture of human relations of domination in their varied degrees. Whilst anthroparchal relations broadly constitute the social relations of nature, natural phenomena exert their own properties and powers in specific situations. Some parts of the life world may not experience in any significant way the effects of such dominations. Yet the term "human domination" does not imply that all humans, are in dominatory relations to the environment, or that all humans continually engage in exploitative and oppressive practices. Social and economic location and the interpenetration of cross cutting structures of various systems of domination mean that some groups of us are positioned in more potentially exploitative relations than others. Some structures of anthroparchy may have intensified in degree whilst others may have lessened over time. I suggested five structures that intermesh to form a social system that we might call anthroparchy. They include anthroparchal relations in production, domestication, violence, culture and the state.

Anthroparchal relations in production

The production relations of anthroparchy have long involved the use of "nature" as a series of resources for the satisfaction of human ends. Production is a crucial link between humanity and "the environment" for humans interact with nature in order to survive. The mass production of goods and services associated with modernity and the industrialization of production significantly increased the ecological footprints of certain groups of humans. The globalizing tendencies of Western industrial practices and process has led to industrialized production being a dominant structure shaping environment–society relations across much of the globe. Technologies and institutions of industrial modes of production are tightly interwoven with the drive for profit maximization and the division of labor. Capitalism, as a system of exploitative social relations also implies the commodification of nature as products, resources and waste. Whilst the global domination of capitalism has never been so deep and diversified however, the nature of production relations is shifting in certain important ways, and the financial networks of contemporary international capital are inherently unstable. Nevertheless a geography of exclusion affects vast regions of the globe, and can be seen embedded in prosperous localities also. Structures of productive exclusion and the dominatory power of capital has led to the disruption of the society/environment spanning ecosystems, particularly of poorer regions of the globe.

Anthroparchal reproduction and domestication

Systemic domination, being dynamic and transformatory, shifts over time and new structures may emerge, or the relative significance of various structures may alter. Anthroparchal innovation has characterized human engagements with the environment for millennia, through the breeding of (hybrid) plants for crops and animals for food and labor.

The last two centuries of Western development have seen a problematic intensification of such processes through industrialized reproduction of plants and animals. Less directly, the reproductive systems of certain species are affected by human interventions such as commercial fishing, wherein the population structure of a species in a particular time and place shifts dramatically. Ecofeminists have noted the gendering of such reproductive interventions, particularly with respect to animal domestication. Such domestication may involve physical confinement, the appropriation of labor and fertility and entrapment. It may also operate at a symbolic level, for example, in the "need" to civilize and "tame" a wild nature, and the distinction between peoples, species and space, which are safely domesticated, and those that are not. Our heightened capacity for technological intervention in "nature" and the increasingly "artificial" character of reproduction has meant that this structure of human domination has become more significant in the last half century. At the same time however, we have seen important sites of contestation around questions of genetic modification, particularly of food crops. Shifts in anthroparchal structures then may be characterized by progress, regress and stasis at the same time.

Anthroparchal governance

Institutions of governance may reproduce, produce or contest and change relations of domination. States at all levels, local, regional, national and intra national place the interests of humans at the crux of decision making, and can act as direct or indirect agents of anthroparchy. For example, through subsidies for intensive farming, road-building schemes, and destruction of ancient woodland states directly engage in damage to ecosystems. Less directly, states may encourage consumption of natural resources through apparent inaction such as not taxing resources. Yet states can also shift relations and practices of domination, for example, by the inclusion of certain kinds of rights to welfare and even self-determination for certain non-human animals, or placing other boundaries on human relations with the environment that limit our intervention in certain positive ways. Institutions and practices of political power reflect social relations of all our differences in systems of domination, and are an important sphere in which the complexities of overlapping difference are played out.

Violence

I consider symbolic forms of violence, which may suggest physical harm, as forms of violence, in addition to the most usual definition of physical coercion. The definition of violence depends on both culturally specific and "real" notions of subjectivity. Cultural definitions of violence shift over time, place and space. Various ecologisms have debated the ethics of what humans might kill with impunity and what they should not, on grounds of a holist functionalism and an individualist sentiency. For species with greater levels of sentiency, violence can be seen to operate in similar ways to violences affecting humans. A key element of normative definitions of violence is physical damage, and deep and feminist ecologies are right to include the destruction of habitats and eco-systems as a form of violence.

Cultures of exclusive humanism

Anthroparchal culture constructs notions of animality and humanity, culture and nature and other such dichotomies. It encourages high rates of consumption, it may represent nature in multiplicious ways that emphasizes the requisites of human domination, and suggests the forms and practices this might take. The boundaries of "society" are based on the notion of human transcendence from and control over nature. This likewise suggests, the notion of "proper" or civilized humanity has been developed in a provincial Eurocentric context. The interpenetration of social constructions of difference is a key site where various systems of social domination (of gender, of "race," of "nature") intersect.

These structures of anthroparchy are crosscut, in different ways and to different degrees by other formations of difference in systems of social domination. Anthroparchal violence can be seen to involve for example, the destruction of habitat, extinction of species via hunting or the slaughter of animals for meat. Culture can also be seen as anthroparchal, for example, in encouraging resource consumption, legitimating resource depletion and human dominance of other species. Sexuality can also be considered to be an anthroparchal structure, involving material control of the sexuality, fertility and reproduction of animals, and the symbolic feminization and sexualization of human male dominance of the natural environment. The state at all levels is likely to be shaped by a number of oppressive systems: capitalism, racism, and patriarchy, in addition to anthroparchy. It can be seen to have systematic bias toward anthroparchal interests, evidenced in its general conduct and specific policies, but it is also a focus of challenges to human domination and may be a site of remedial action. These structures are most relevant for the analysis of the gendering of "nature": violence, sexuality, culture and the state. Other structures of anthroparchy are not so clearly patriarchal.

Systems of domination change historically in form and degree. I concurred with Walby (1990) that patriarchy has changed and adopted a more public mode of operation, partly through a shift in the relative significance of certain structures: the lessening significance of the household, and increased importance of the public structures of paid employment and the state. I would give increased emphasis to Walby's contention that patriarchal structures have themselves shifted to a more public form. I would argue for example, that this may be seen with respect to: sexuality (such as patriarchal sexualization of popular culture, medicalization of fertility, pregnancy and childbirth), (popular) culture and its current reconfiguration (of masculinities in particular) and the state as a site for both continuing gendered exclusions and for feminist contestation. Anthroparchal structures of dominance can and have also changed, but in most cases, their intensity has increased. For example, violences against animals can be seen to have moved into an increasingly public mode with the advent of the factory farm, and the state, under pressure from contestation by animal welfare activists, places boundaries only on severe malpractice within an already abusive set of institutions and practices. Sexuality can be conceptualized as an all-encompassing domination with the increased application of reproductive technologies. Despite shifts in the content of oppressive structures, and the intensity of their oppressive power, they have some degree of continuity and are flexible and adaptive in re-making themselves.

Multiplicities of power: systemic permeability

A comparison of both systems by examining common and divergent structures is a means we can use to delimit the boundaries of each system. Three of the structures of patriarchy and anthroparchy are common: culture, violence and the state – although their content, form and degree of operation, differs. Structures of systems of oppression can be seen as analytically distinct, but although the three structures are discussed separately, certain elements clearly overlap and interlink.

Culture

The culture of contemporary Western societies can be seen as a structure of both patriarchy and anthroparchy. Patriarchal culture involves the creation and deployment of notions of femininities and masculinities, through various media. For example, despite significant changes in patriarchal structuring, the core of contemporary discourses of femininity is (hetero)sexuality, within which women should be sexually attractive to, and available for, men. Alternatively, there is a discourse of gendered domesticity, according to which women should desire motherhood, and engage in domestic labor, and thereby "care" for their male partner and children. These discourses might be articulated in gendered forms of popular culture including film, "women's" magazines and product advertising. As more women contest and reject this domestic role, the cultural control of women, I would suggest, has shifted toward sexualization of women in popular culture. Anthroparchal culture creates notions of human superiority, and of inferiority of other animals and the natural environment. It also deploys discourses based upon sexualization and domestication. In this case, domestication does not involve the notion of service. The environment is symbolized as a dangerous wilderness that must be subject to domination, as can be seen in the extensive and sometimes unnecessary cultivation of land, the "management" of forests and domestication, and killing or regulation of wild animals. The environment is constructed as a series of objects over which humans may exercise control. Anthroparchal culture encourages consumption and the value of "affluence," which obscures and legitimates resource depletion and certain violences. These discourses of consumption and domestication may operate in tandem with the sexualization of human dominance, in which environmental control is constructed as sexual domination (such as cultivation of "virgin" territory). The Enlightenment prioritization of human reason legitimated both animal abuse, and exploitation of the environment for human benefit in modernity. Some eco-feminists have rightly noted such discourse was also part of patriarchal culture that constructed women as less human. The cultural structures of patriarchy and anthroparchy shape behavior, and justify differences that are hierarchically conceived. Whilst discourses evident in popular culture are important in the construction and maintenance of patriarchal and anthroparchal power relations, these do not go uncontested, and culture is a site in which complex differences coalesce and compete, and in which social movements re/shape and shift narratives of domination.

Sex, sexuality and reproduction

Sexuality is constituted through multiple discourses based on power relations of gender and nature. Discourses of gendered and natured sexuality often interlink, for example, the "Othering" of patriarchal sexualization may not only involve feminization according to gendered domination, but animalization according to natured domination. Radical feminists have seen sexuality as a key structure of patriarchy, involving normative heterosexuality, sexual harassment, and the general sexualization of gender domination. I concur with those arguing that sexuality changed form during the twentieth century, adopting an increasingly public mode with the decline of privatized sexual control of women in the household. If we define sexuality as encompassing fertility and reproduction, comparisons may be drawn with anthroparchy. For example, domesticated animals that are "livestock" have no control over their sexuality and fertility. The basis of domestication for meat production lies in reproductive and sexual control. Reproductive technologies were developed in the context of meat and dairy farming, with the purpose of increasing human control of fertility and reproduction in animals. Technologies to control animal reproduction are increasingly applied to women with the medicalization of fertility, pregnancy and birth. Anthroparchal constructions of sexuality involve absolute domination of animals, whereas patriarchal sexuality is contested through human agency, and in certain aspects, patriarchal sexuality has altered. Anthroparchal sexuality can be seen in the discursive sexualization of the control of the environment, sexualization of the control of animals and their domestication, or the characterization of particular forms of animal abuse as sexual.

Violence

Violence may take symbolic form for images may for example, recall actual physical violence, but it overwhelmingly assumes physically coercive form. At the material level of physical violence, we see the greatest difference between the systems of domination, because the forms structures of violence assume differ between patriarchy and anthroparchy, and so do the degrees at which their practices operate. Violence is endemic in patriarchy and anthroparchy, but it operates to a lesser degree in patriarchy (i.e., it is less systematic, and less extreme) and adopts a wide variety of different forms (slaughter, destruction of forest eco-systems in one case, sexual harassment in the other, for example). Patriarchal violence is seen by feminists as constituting violence against women (and children); appearing in a number of forms and having differing degrees of severity: from rape, child sexual abuse, the battering of female partners to the less physically harmful instances of sexual harassment. Women's behavior may be restricted because of fear of violence, and violence may exist where its presence is suggested, for example in certain pornographic images. Whilst these are instances of violence, they represent lesser degrees than forms of physical violence. Anthroparchal violence is likely to adopt physical form and operate to an intense degree. The type of violence differs in relation to the aspect of the environment it affects, and the degree of sentiency possessed by certain communities of organisms. Anthroparchal violence can be indirect, such as destruction of habitat that results in species extinction, but natured violence operates to extreme degrees in terms of human treatment of some Other animals. In Western high modernity, for example, most domestic animals live on farms within industrial systems of control deriving from their status as food production units. Animals are often incarcerated, may experience battery, rape, electric shock, and slaughter and suffer psychological abuse, such as separation from kin and peers.

The state and governance

Governance is shaped by structural considerations pertaining to various systems of social domination. The gendering and naturing of the state can be evidenced as much in what it does not do, as in its decisions and policy, for certain issues, grievances and constituencies may be excluded from policy making and other forms of consideration. The state may intervene to positively support structures of, for example, concealing information about diseases in food production from a concerned public, in order to support intensive animal farming, or the gendering of benefits and employment leave for new parents. Alternatively, the state may support oppressive systems via its non-intervention, which functions to protect and maintain such systems, such as the decision not to tax industrial waste. In recent decades, largely in response to feminist political action, the state has shifted policy regarding gender relations, resulting in some benefits for women (decriminalization of abortion, equal opportunities legislation), and particularly at the local and regional government level, feminism has been seen to have some impact on policy making. However, certain other policies have, in an indirect way, had negative effects on women (such as cuts in welfare provision that disadvantage women as primary carers). The role of the state in reinforcing gender relations can be seen largely in its lack of intervention to protect women and act against inequalities that legitimate the patriarchal status quo. Similarly, the state may act to limit excesses of violence against animals, or the use of particularly harmful chemicals in agricultural production. By such action to ameliorate excesses and particular abuses, conditions for the operation of the anthroparchal institutions and practices are maintained in the longer term. The power of the state at various levels of governance makes it an important site for the contestation of relations of domination and the playing out of the complexities of difference. Whilst all the structural elements of a system interrelate and impinge on others, the state has a privileged role in effecting patterns of change within a social system as a whole, as it has the power, incrementally in the main, to enforce changes in normative practice in social institutions through the power of law.

Structures of distinction

The household is a site of privatized production relations and is of less significance in the control of women in the twenty-first century, than in the past. It can still be considered a patriarchal structure due to the continuing gendered division of domestic labor; domestic violence against women, and the ways women's domestic labor may impinge upon their position in paid employment. It is not a structure of anthroparchy. Although the household is an important site of mass consumption that affects the environment, this consumption is more accurately seen as part of the anthroparchal structure of industrialization. Some animals are kept in households as "pets," but this is an aspect of anthroparchal domestication. There are some similarities between the domestication of animals and of women within the household in the forms these oppressive relations adopt such as physical confinement and the appropriation of labor, but I feel there is sufficient difference in the content and form of anthroparchal domestication and the patriarchal household, to conceptualize them separately. Paid employment should likewise be seen as a patriarchal and not an anthroparchal structure. Despite significant legislative changes and shifts in cultural normatives, and despite local variations within and across wealthy regions of the globe, women in patriarchal society are paid less than men and horizontally and vertically segregated in low status employment that is clearly gendered. This gendering of paid employment varies locally, nationally and in terms of global patterns of relative inclusion and exclusion. There is a distinct gendered division of labor in which women are segregated in certain parts of the job market, and at certain levels within it. The specific type of work and the particular pattern of that structuring is historically and spatially variable.

I use the term "domestication" to refer to certain exercises and institutional formations of human domination. Anthroparchal domestication may involve the management and control of the wilderness, the cultivation of land, and use of land for rearing animals that have been rendered docile by behavioral modification and genetic manipulation in order that they become human resources. Whilst there is probably

Domination in a Lifeworld of Complexity 175

a gendered element to such domestication, such as the feminization of the land/animals domesticated, the basis of domestication is natured difference, and the domestication of nature may have a different purpose to the domestic role of women in the household. The former is anthroparchally necessitated to control a wilderness, which is constructed as potentially dangerous for humans, whereas the latter is necessitated by capitalism and patriarchy for the expropriation of female labor. Finally, whilst the patriarchal household is largely a privatized structure, anthroparchal domestication is public. Household domestication is primarily an individual form of appropriation. Anthroparchal domestication involves collective appropriation of the labor and bodies of animals in mass industries to produce commodities for mass consumption, and the appropriation of the bodies and forms of plants, for example.

Where patriarchy and anthroparchy have structures in common, it is likely that there may be close relationships between the systems of oppression, or a high level of "interpenetration." However, the content of patriarchal and anthroparchal structures, even when common, is divergent in form and degree. Relationships between systems of oppression would be characterized by tension and conflict, as well as co-operation and mutual accommodation, or co-adaptation. Systems of domination based on difference interrelate in complex and contradictory ways and are best conceived as independent whilst also interrelated. Ecological feminism has already attempted to conceptualize these interrelations between different kinds of domination in terms of overlapping "spheres," overarching "logic"(s), congruent webs, often involving capitalism, patriarchy and postcoloniality in addition to the domination of nature, which are sometimes seen as a single all-encompassing system of domination.

My own conception is more fractured. Structures of systems of domination are the nexus in which we can see intermeshing and overlapping relations of power. Elements of structures may not coalesce in utterly similar ways; there will be different articulations of overlap and similarity of institutional organization and practice. The elements of articulation may demonstrate unique formations and properties in specific locations (which are spatialized and timealized). The extensive differences in form and degree of domination exerted by different systems of domination in different times, places and spaces, means, in my view, that a multiple systems approach enables us to theorize the complexities of structural dynamics in a matrix of domination, without marginalizing difference.

The problem of difference revisited – life beyond the matrix?

This book has avoided discussion of pathways out of the matrix of domination, and this may lead one to feel it is rather pessimistic, perhaps even defeatist. Pierre Bourdieu claimed, like Luhmann suggests, that social stasis is more prevalent than social change in justifying what critics (see Lash 1990) have held to be the structural(ist) focus of his work. In many ways I concur with Bourdieu, but I also consider that substantive change is often less apparent than it might be, given its impact within webs of domination. I would endorse Sandra Barkty's defense here when she suggests that:

> Theoretical work done in the service of political ends may exhibit a "pessimism of the intellect," but the point of doing such work at all is that "optimism of the will" without which any serious political commitment is impossible. (1990:7)

To take on the task of "what is to be done" is both gargantuan and probably impossible. Political action is emergent in particular locations and contexts, although it is also usually informed by a political understanding and perhaps some varied theoretical positions on the part of its adherents. Whilst I would like to see a political understanding and commitment informed by the mutiplicitous and complex qualities of difference in domination, a multiple systems approach to understanding the complexity of domination cannot specify political choices. What I would say however, is that we require a radical pluralist toleration of difference in contesting the intermeshing of dominations, and that we might struggle to see difference as diversity and not as a matter of hierarchical relations.

The richness of diversity

Complexity theory assumes a relationship between order and chaos. In each system, social or natural, this relationship will play itself out in a unique and specific way. Systems bifurcate – they choose (and this is a *metaphor*, let it be noted!) between possible paths to becoming. Systems are not teleological – we cannot predict the paths adopted, and we are likely to be surprised by patterns of change in social systems, captured in the blink of a human lifetime. The changes we see in social systems are dependent on the systems history, and on its interrelation with other systems and other external conditions. The scientific theory of dissipative structures indicated that systems are not equilibrist, but operate far from universal laws. As we move from equilibrium, we move toward diversity and away from unity, toward enhanced variety.

I have argued that social systems are not teleological and are dynamic, patterned in complexity, often apparently chaotic. Social systems of domination seek to impose regularized formations, structures of normativity on both the multivariate of human difference, and on the even more variant ways in which human communities are diversely implicated in natural relations. In seeking to capture such tendencies toward normativity, theories of patriarchy and capitalism, for example, have fallen foul, it is said, of an appreciation of diversity and difference.

The state, for example, is constantly a focus for challenges to gendered and natured power, and of other dominations. Institutions of global governance such as the United Nations and the institutions of capitalist regulation, such as the World Trade Organization have come under significant pressure from those defending human rights and agitating for environmental protection. The European Union is pressurized from within and without by an agenda of state feminism which agitates for a "Scandinavian model" of welfare provision and a work culture which is cognizant of time as a resource which is gendered and which manages work practices in order to ameliorate the effects of the labor market. Such a model, may allow for workers to engage with children, companion animals and rich lives outside of work and in work. Environmental justice organizations agitate in both local and global forums for an understanding that specific communities incur particular environmental "bads." To list the challenges of feminism and ecologism, to indicate the links in political praxis which both have made, with actions which contest gendered, natured, capitalist and postcolonial relations of power would take at least another book.

The complexity of systems, natural and social, indicates that diversity is a strength. So too, is it likely to be in an attempt to unpick the matrix of social domination. Protecting endangered species and their habitats, agitating for animal welfare, are not bourgeois pastimes, particularly when those species and habitats are endangered by the strategic actions of multi-billion corporations. The anti-globalization "movement," disparate and fractured, differentially operative though it may be, is an example of political activism which connects the systemic operation of capitalism to the abuse of both the environment and peoples, particularly in poorer regions of the globe. Those of us who adhere to any kind of radically contestationary politics and oppose the dominations of race, of class, of gender or of nature or of other exclusions, exploitations, oppressions, need to look hard for political links, although our own lifeworlds and localities will often specify those choices for us. Much as change may be a struggle, it is also so often embedded. We might not notice as our practices of daily life remake social structures through our sexual relations, kin relations, household formations, employment choice and practices, consumption patterns, use of "leisure" space and time. Incredible and dramatic change is much needed, is multifaceted, complex, complicated. It may all seem too much, too improbable and impossible. But potentialities for remaking our relations are embedded in the very detail of the matrix of domination.

Bibliography

Adam, B. 1998, *Timescapes of Modernity: The Environment and Invisible Hazards*. London: Routledge.

Adam, B., Beck, U. and Van Loon, J. 2000, *The Risk Society and Beyond: Critical Issues for Social Theory*. London: Sage.

Adam, B. and Van Loon, J. 2000, "Repositioning Risk: The Change for Social Theory," in B. Adam, U. Beck and J. Van Loon. eds. *The Risk Society and Beyond: Critical Issues for Social Theory*. London: Sage.

Adams, C. J. 1976, "The Inedible Complex: The Political Implications of Vegetarianism," *Second Wave* 4, 1:36–42.

Adams, C. J. 1990, *The Sexual Politics of Meat*. Cambridge: Polity.

Adams, C. J. 1994, *Neither Man nor Beast: Feminism and the Defense of Animals*. New York: Continuum.

Adams, C. J. 1995, "Woman Battering and Harm to Animals," in C. J. Adams and J. Donnovan. eds. *Animals and Women: Feminist Theoretical Explorations*. London: Duke University Press.

Adams, C. J. and Donovan, J. eds, 1995, *Animals and Women: Feminist Theoretical Explorations*. London: Duke University Press.

Adams, C. J. 2003, *The Pornography of Meat*. London: Continuum.

Adkins, L. and Merchant, V. 1996, *Sexualizing the Social: Power and the Organization of Sexuality*. Explorations in Sociology 47. London: Macmillan.

Adler, M. 1986, *Drawing Down the Moon: Witches, Goddess-Worshippers and Other Pagans in America Today*. Boston, MA: Beacon.

Afshar, H. and Maynard, M. 1994, *Dynamics of "Race" and Gender: Some Feminist Interventions*. London: Taylor and Francis.

Agarwal, B. 1986, *Cold Hearths and Barren Slopes: The Woodfuel Crisis in the Third World*. London: Zed Books.

Agarwal, B. 1992, "The Gender and Environment Debate: Lessons from India," *Feminist Studies* 18, 1:119–58.

Alaimo, S. 1994, "Cyborg and Ecofeminist Interventions: Challenges for an Environmental Feminism," *Feminist Studies* 20,1:133–52.

Albrecht, G. A. 1998, "Ethics and Directionality in Nature," in A. Light. ed., *Social Ecology After Bookchin*. New York: The Guilford Press.

Albrow, M. 1996, *The Global Age*. Cambridge: Polity.

Alcoff, L. 1988, "Cultural Feminism versus Post-Structuralism: The Identity Crisis in Feminist Theory," *Signs* 13, 3:405–36.

Alcoff, L. and Potter, E. eds, 1993, *Feminist Epistemologies*. London: Routledge.

Alexander, S. and Taylor, B. 1980, "In Defence of Patriarchy," in M. Evans. ed., 1982, *The Woman Question*. London: Fontana.

Almond, B. 1995, "Rights and Justice in the Environment Debate," in D. Cooper and J. Palmer. eds. *Just Environments: Intergenerational, International and Interspecies Issues*. London: Routledge.

Althusser, L. 1963, "On the Materialist Dialectic," in L. Althusser, *For Marx*. London: Allen Lane.

Bibliography

Althusser, L. 1968, "The Object of Capital," in L. Althusser and E. Balibar. eds, *Reading Capital*. London: New Left Books.
Amin, S. 1997, *Capitalism in the Age of Globalization*. London: Zed.
Amos, V. and Parmar, P. 1984, "Challenging Imperial Feminism," *Feminist Review* 17:3–20.
Anderson, K. 2001, "The Nature of 'Race'," in N. Castree, and B. Braun. eds, *Social Nature*. Oxford: Blackwell.
Antonio, D. 1995, "Of Wolves and Women," in C. J. Adams and J. Donovan. eds, *op cit*.
Arac, J. ed. 1988, *After Foucault: Humanistic Knowledge, Postmodern Challenges*. London: Rutgers University Press.
Archer, M. S. 1995, *Realist Social Theory: The Morphogenetic Approach*. Cambridge: Cambridge University Press.
Archer, M. S. 1996, "Social Integration and System Integration: Developing the Distinction," *Sociology* 30, 4:679–99.
Arditti, R., Duelli Klein, R. and Minden, S. eds, 1984, *Test Tube Women*. London: Pandora Press.
Arnold, D. 1996, *The Problem of Nature: Environment, Culture and European Expansion*. Oxford: Blackwell.
Assiter, A. 1994, *Enlightened Women: Modernist Feminism in a Postmodern Age*. London: Routledge.
Attfield, R. 1983, *The Ethics of Environmental Concern*. Oxford: Blackwell.
Aziz, R. 1997, "Feminism and the Challenge of Racism: Deviance or Difference," in H. Mirza. ed., *Black British Feminism: A Reader*. London: Routledge.
Bacchi, C. 1990, *Same Difference: Feminism and Sexual Difference*. London: Allen and Unwin.
Bahro, R. 1982, *Socialism and Survival*. London: Heretic Books.
Bahro, R. 1984, *From Red to Green: Interview with New Left Review*. London: Verso.
Bahro, R. 1986, *Building the Green Movement*. London: GMP.
Bailey, M. E. 1993, "Foucauldian Feminism: Contesting Bodies, Sexuality and Identity," in C. Ramazanoglu. ed., *Up Against Foucault*. London: Routledge.
Baker, P. L. 1993, "Chaos, Order and Sociological Theory," *Sociological Inquiry* 63, 4:406–24.
Balsamo, A. 1996, *Technologies of the Gendered Body: Reading Cyborg Women*. London: Duke University Press.
Barbre, J. W. 2003, "Mono-Boomers and Moral Guardians: An Exploration of the Cultural Construction of the Menopause," in R. Weitz. ed., *The Politics of Women's Bodies: Sexuality, Appearance and Behavior*. Second edition. Oxford: Oxford University Press.
Barrett, M. 1980, *Women's Oppression Today*. London: Verso.
Barrett, M. and McIntosh, M. 1982, *The Anti-Social Family*. London: Verso.
Barrett, M. and McIntosh, M. 1985, "Ethnocentrism and Socialist Feminist Theory," *Feminist Review* 20, 22–42.
Barrett, M. 1987, "The Concept of Difference," *Feminist Review* 26:29–41.
Barrett, M. and Philips, A. 1992, *Destabilizing Theory Contemporary Feminist Debates*. Cambridge: Polity Press.
Barry, J. 1999, *Environment and Social Theory*. London: Routledge.
Barry, K. 1996, "Deconstructing Deconstructionism," in D. Bell and R. Klein. eds, *Radically Speaking*. London: Zed Books.

Bartky, S. 1990, *Femininity and Domination: Studies in the Phenomenology of Oppression*. "Introduction." London: Routledge.
Bartky, S. 1990, "Feeding Egos and Tending Wounds: Deference and Disaffection in Women's Emotional Labor," in S. Bartky, *Femininity and Domination: Studies in the Phenomenology of Oppression*. London: Routledge.
Bartky, S. 1988, "Foucault, Femininity and the Modernization of Patriarchal Power," in Diamond and L. Quinby. eds, *Feminism and Foucault: Reflections on Resistance*. Boston: Northeastern University Press.
Bartky, S. 1982, "Narcissism, Femininity and Alienation," in S. Bartky, 1990 *Femininity and Domination: Studies in the Phenomenology of Oppression*. London: Routledge.
Bartky, S. 1979, "On Psychological Oppression," in S. Bartky, 1990, *Femininity and Domination: Studies in the Phenomenology of Oppression*. London: Routledge.
Bartky, S. 1976, "Towards a Phenomenology of Feminist Consciousness," in S. Bartky. ed., *Femininity and Domination: Studies in the Phenomenology of Oppression*. London: Routledge.
Baudrillard, J. 1983, *Simulations*. New York: Semiotexte.
Bauman, Z. 1991, *Modernity and Ambivalence*. Cambridge: Polity.
Beck, U. 1992, *The Risk Society: Towards a New Modernity*. London: Sage.
Beck, U. 1996, "Risk Society and the Provident State," in S. Lash, B. Szerszynski and B. Wynne. eds. *Risk, Environment and Modernity*. London: Sage.
Beck, U. 1995, *Ecological Politics in an Age of Risk*. Cambridge: Polity.
Beck, U. 1999, *World Risk Society*. Cambridge: Polity.
Beck, U. 2000, "Risk Society Revisited," in B. Adam, U. Beck, and J. van Loon. eds. *The Risk Society and Beyond: Critical Issues for Social Theory*. London: Sage.
Beder, S. 1997, *Global Spin: The Corporate Assault on Environmentalism*. Dartington: Green Books.
Beechey, V. 1979, "On Patriarchy," *Feminist Review* 3:66–82.
Beechey, V. 1978, "Women and Production: A Critical Analysis of Some Sociological Theories of Women's Work," in A. Khun and A. M. Wolpe. eds, *Feminism and Materialism: Women and Modes of Production*. London: Routledge.
Beechey, V. 1987, *Unequal Work*. London: Verso.
Bell, D. and Klein, R. eds, 1996, *Radically Speaking: Feminism*. London: Zed Books.
Bell, M. M. 1998, *An Invitation to Environmental Sociology*. Thousand Oaks, CA: Pine Forge Press.
Benhabib, S. 1992, *Situating the Self: Gender, Community and Postmodernism in Contemporary Ethics*. Cambridge: Polity.
Benholt Thompsen, V. and Mies, M. 1999, *The Subsistence Perspective: Beyond the Globalized Economy*. London: Zed.
Benny, N. 1983, "All One Flesh: The Rights of Animals," in L. Caldecott and S. Leyland. eds, *Reclaim The Earth*. London: Women's Press.
Benton, T. 1981, "Realism in Social Science," *Radical Philosophy* 27: 13–21.
Benton, T. 1984, *The Rise and Fall of Structural Marxism*. Basingstoke: Macmillan.
Benton, T. 1985, "Realism and Social Science," in R. Edgley and R. Osborne. eds, *Radical Philosophy Reader*. London: Verso.
Benton, T. 1988, "Marx, Humanism and Speciesism," *Radical Philosophy* 50: 4–18.
Benton, T. 1989, "Marxism and Natural Limits: An Ecological Critique and Reconstruction," *New Left Review* 178: 51–86.

Benton, T. 1991, "Biology and Social Science: Why the Return of the Repressed should be Given a Cautious Welcome," *Sociology* 25,1:1–30.
Benton, T. 1992, "Why the Welcome Needs to be Cautious: A Reply to Keith Sharpe," *Sociology* 26, 2:225–33.
Benton, T. 1993, *Natural Relations: Ecology, Animal Rights and Social Justice*. London: Routledge.
Benton, T. 1994, "Biology and Social Theory in the Environment Debate" in M. Redclift, and T. Benton. eds, *Social Theory and the Global Environment*. London: Routledge.
Benton, T. and Redclift, M. 1994, Introduction, in *ibid*.
Benton, T. ed., 1996, *The Greening of Marxism*. London: Guilford.
Benton, T. 1998, "Why are Sociologists Nature-Phobes?" paper to the Centre for Critical Realism Conference – After Postmodernism: Critical Realism? University of Essex.
Bernauer, J. and Rasmussen, D. eds, 1988, *The Final Foucault*. Cambridge: MIT Press.
Bhaskar, R. 1978, *A Realist Theory of Science*. Second edition. Brighton: Harvester Press.
Bhaskar, R. 1979, *The Possibility of Naturalism*. Second edition. Brighton: Harvester Press.
Bhaskar, R. 1989, *Reclaiming Reality*. London: Verso.
Biehl, J. 1988, "What is Social Ecofeminism?" *Green Perspectives* 11:1–8.
Biehl, J. 1991, *Rethinking Ecofeminist Politics*. Boston, MA: South End Press.
Birke, L. 1986, *Women, Feminism and Science: The Feminist Challenge*. Brighton: Harvester.
Birke, L. 1991, "Science, Feminism and Animal Natures I: Extending the Boundaries," *Women's Studies International Forum* 14, 5:443–50.
Birke, L. 1992, "In Pursuit of Difference," in L. Keller and G. Kirkup. eds, *Inventing Women*. Cambridge: Polity.
Birke, L. 1994, *Feminism, Animals and Science: The Naming of the Shrew*. Buckingham: OUP.
Birke, L. 1995, "Exploring the Boundaries: Feminism, Animals and Science," in C. J. Adams, and J. Donovan. eds, *op cit*.
Birke, L. 1999, *Feminism and the Biological Body*. Edinburgh: Edinburgh University Press.
Birke, L. and Hubbard, R. eds, 1995, *Reinventing Biology: Respect for Life and the Creation of Knowledge*. Bloomington, IL: Indiana University Press.
Blaikie, P. 2001, "Social Nature and Environmental Policy in the South," in N. Castree and B. Braun. eds, 2001, *Social Nature*. Oxford: Blackwell.
Blaikie, P. and Brookfield, H. 1987, *Land Degradation and Society*. London: Methuen.
Bleier, R. 1984, *Science and Gender*. Elmsford, NY: Pergammon.
Bloor, D. 1999, "Anti-Latour," *Studies in the History and Philosophy of Science* 30:81–112.
Boden, D. 2000, "Worlds in Action: Information, Instantaneity and Global Futures Trading," in U. Beck, "Risk Society Revisited," in B. Adam, U. Beck, and J. van Loon, J. eds, *The Risk Society and Beyond: Critical Issues for Social Theory*. London: Sage.
Boerner, C. and Lambert, T. 1995, "Environmental Injustice," *The Public Interest* 95, 118:61–82.

Bookchin, M. 1971, *Post-Scarcity Anarchism*. Berkeley, CA: Ramparts Press.
Bookchin, M. 1980, *Toward an Ecological Society*. Montreal: Black Rose Books.
Bookchin, M. 1986, *The Modern Crisis*. Philadelphia, PA: New Society.
Bookchin, M. 1989, *Remaking Society*. Montreal: Black Rose Books.
Bookchin, M. 1990, *The Philosophy of Social Ecology*. Montreal: Black Rose Books.
Bookchin, M. 1991, *The Ecology of Freedom*. Montreal: Black Rose Books.
Bookchin, M. 1992, *Urbanization Without Cities*. Montreal: Black Rose Books.
Bookchin, M. 1995, *Re-Enchanting Humanity*. London: Cassells.
Bookchin, M. 1995, *Social Anarchism or Lifestyle Anarchism*. San Francisco, CA: AK Press.
Bordo, S. 1990, "Feminism, Postmodernism, and Gender Skepticism," in L. Nicholson. ed., *Feminism/Postmodernism*. London: Routledge.
Bordo, S. 1993, "Feminism, Foucault and the Politics of the Body," in C. Ramazanoglu. ed., *Up Against Foucault*. London: Routledge.
Bottero, W. 1998, "Clinging to the Wreckage? Gender and the Legacy of Class," *Sociology* 32, 3:469–90.
Bourdieu, P. 1984, *Distinction: A Social Critique of the Judgement of Taste* trans. R. Nice, R.Cambridge, Mass.: Harvard University Press.
Bowen, A. 1996, "Enabling a Visible Black Lesbian Presence in Academia," in D. Bell and R. Klein. eds, *Radically Speaking*. London: Zed Books.
Bowles, G. and Duelli Klein, R. 1983, *Theories of Women's Studies*. London: Routledge.
Boyne, R. and Rattansi, A. eds, 1990, *Postmodernism and Society*. London: Macmillan.
Bradford, G. 1989, *How Deep is Deep Ecology?* Ojai: Times Change Press.
Bradley, H. 1996, *Fractured Identities: Changing Patterns of Inequality*. Cambridge: Polity.
Brah, A. 1996, *Cartographies of Diaspora*. London: Routledge.
Braidotti, R. 1989, "Organs Without Bodies," *Differences* 1:147–61.
Braidotti, R. 1991, *Patterns of Dissonance*. Cambridge: Polity.
Braidotti, R. 2002, *Metamorphoses: Towards a Materialist Feminist Theory of Becoming*. Cambridge: Polity.
Braidotti, R., Charkiewicz, E., Havscher, S. and Wieringa, S. 1994, *Women, the Environment and Sustainable Development*. London: Zed Books.
Braidotti, R. and Lykke, N. 1996, *Between Monsters, Goddesses and Cyborgs*. London: Zed Books.
Braun, B. and Castree, N. eds, 1998, *Remaking Reality: Nature at the Millennium*. London: Routledge.
Braun, B. and Wainwright, J. 2001, "Nature, Poststructuralism and Politics," in N. Castree, N. and B. Braun, B. eds, *Social Nature*. Oxford: Blackwell.
Braverman, H. 1974, *Labour and Monopoly Capital*. New York: Monthly Review Press.
Brenner, J. 2000, *Women and the Politics of Class*. New York: Monthly Review Press.
Breugel, I. 1979, "Women as a Reserve Army of Labour," *Feminist Review* 3:12–23.
Brod, H. 1994, *Theorizing Masculinities*. London: Sage.
Brodribb, S. 1992, *Nothing Matters: A Feminist Critique of Postmodernism*. New York: New York University Press.
Brodribb, S. 1996, "Nothing Matters" in D. Bell and R. Klein. eds, *Radically Speaking*. London: Zed Books.

Brownmiller, S. 1976, *Against Our Will*. Harmondsworth: Penguin.
Brownmiller, S. 2000, *In Our Time: Memoir of a Revolution*. London: Aurum Press.
Bryant, R. L. 2001, "Political Ecology: A Critical Agenda for Change?" in N. Castree and B. Braun. eds, *Social Nature*. Oxford: Blackwell.
Bryld, M. and Lykke, N. 1999, *Cosmodolphins: Feminist Cultural Studies of Technology, Animals and the Sacred*. London: Zed.
Bryson, V. 2003, *Feminist Political Theory*. Second Edition. Basingstoke: Palgrave.
Buege, D. J. 1994, "Rethinking Again: A Defense of Ecofeminist Philosophy," in K. J. Warren. ed., *Ecological Feminism*. London: Routledge.
Bullard, R. 1990, *Dumping in Dixie: Race, Class and Environmental Quality*. Boulder, CO: Westview Press.
Bullard, R. 1993, *Confronting Environmental Racism: Voices from the Grassroots*. Boston, MA: South End Press.
Bunyard, P. and Morgan-Grenville, F. eds, 1987, *The Green Alternative*. London: Methuen.
Burningham, K. and Cooper, G. 1999, "Being Constructive: Social Constructionism and the Environment," *Sociology* 33, 2: 297–316.
Butler, J. 1990, *Gender Trouble: Feminism and the Subversion of Identity*. London: Routledge.
Butler, J. 1993, *Bodies that Matter: On the Discursive Limits of "Sex."* London: Routledge.
Butler, J. 1997, *Excitable Speech: A Politics of the Performative*. London: Routledge.
Butler, J. 1999, "Preface 1999" in *Gender Trouble: Feminism and the Subversion of Identity*. London: Routledge.
Butler, J. and Scott, J. eds, 1992, *Feminists Theorize the Political*. London: Routledge.
Buttel, F. 1997, "Social Institutions and Environmental Change," in M. Redclift and G. Woodgate. eds, *The International Handbook of Environmental Sociology*. Cheltenham: Edward Elgar.
Byrne, D. 1998, *Complexity Theory and the Social Sciences*. London: Routledge.
Caldecott, L. and Leyland, S. eds, 1983, *Reclaim the Earth*. London: Women's Press.
Callinicos, A. 1989, *Against Postmodernism: A Marxist Critique*. Cambridge: Polity.
Callinicos, A., Rees, J., Harman, C. and Haynes, M. 1994, *Marxism and the New Imperialism*. London: Bookmarks.
Capra, F. 1976, *The Tao of Physics*. London: Flamingo.
Capra, F. 1983, *The Turning Point*. London: Fontana.
Capra, F. 1996, *The Web of Life: A New Synthesis of Mind and Matter*. London: HarperCollins.
Capra, F. 2003, *The Hidden Connections: A Science for Sustainable Living*. London: Flamingo.
Capra, F. and Steindl-Rast, D. 1991, *Belonging to the Universe*. San Francisco: Harper and Row.
Caputi, J. 1989, *The Age of Sex Crime*. London: Women's Press.
Carby, H. 1982, "White Women Listen! Black Feminism and the Boundaries of Sisterhood," in Centre for Contemporary Cultural Studies, University of Birmingham, *The Empire Strikes Back: Race and Racism in '70's Britain*. London: Hutchinson.
Castells, M. 1996, *The Information Age Vol. 1: The Rise of the Network Society*. Oxford: Blackwell.

Castells, M. 1998, *The Information Age Vol. 3: The End of the Millenium*. Oxford: Blackwell.
Castells, M. 2000, "Information Technology and Global Capitalism," in W. Hutton. and A. Giddens. eds, *Global Capitalism*. New York: The New Press.
Castree, N. 1995, "The Nature of Produced Nature," *Antipode* 27:12–47.
Castree, N. 2001, "Socializing Nature: Theory, Practice and Politics," in N. Castree and B. Braun, eds, *Social Nature*. Oxford: Blackwell.
Castree, N. 2001, "Marxism, Capitalism and the Production of Nature," in N. Castree and B. Braun. eds, *Social Nature*. Oxford: Blackwell.
Castree, N. and MacMillan, T. 2001, "Dissolving Dualisms: Actor-Networks and the Reimagination of Nature," in N. Castree and B. Braun. eds, *Social Nature*. Oxford: Blackwell.
Castree, N. and Braun, B. eds, 2001, *Social Nature*. Oxford: Blackwell.
Catton, W. R. and Dunlap, R. E. 1978, "Environmental Sociology: A New Paradigm," *The American Sociologist* 13:41–9.
Catton, W. R. and Dunlap, R. E. 1980, "A new Paradigm for Post-Exuberent Sociology," *American Behavioral Scientist* 24, 1:15–47.
Charles, N. and Kerr, M. 1988, *Women, Food and Families*. Manchester: Manchester University Press.
Charles, N. and Hughes-Freeland, F. 1996, *Practising Feminism: Identity, Difference, Power*. London: Routledge.
Chase, B. ed., 1991, *Defending the Earth: A Dialogue Between Murray Bookchin and Dave Foreman*. Boston: South End Press.
Chatterjee, P. and Finger, M. 1994, *The Earth Brokers: Power, Politics and World Development*. London: Routledge.
Cheney, J. 1994, "Nature/Theory/Difference: Ecofeminism and the Reconstruction of Environmental Ethics," in K. J. Warren. ed., *Ecological Feminism*. London: Routledge.
Chesler, P. 1978, *About Men*. London: Women's Press.
Chisholm, D. 1995, "The 'Cunning Lingua' of Desire: Bodies-Language and Perverse Performativity," in E. Grosz and E. Probyn. eds, *Sexy Bodies: The Strange Carnalities of Feminism*. London: Routledge.
Chodorow, N. 1978, *The Reproduction of Mothering*. Berkeley, CA: University of California Press.
Christ, C. P. and Plaskow, J. eds, 1989, *Weaving the Visions: New Patterns in Feminist Spirituality*. San Francisco, CA: Harper and Row.
Christ, C. P. 1992, "Spiritual Quest and Women's Experience," in C. P. Christ and J. Plaskow. eds, *Womanspirit Rising*. San Francisco: HarperCollins.
Christ, C. P. and Plaskow, J. eds, 1992, *Womanspirit Rising*. San Francisco, CA: Harper and Row.
Cilliers, P. 1998, *Complexity and Postmodernism: Understanding Complex Systems*. London: Routledge.
Clark, J. 1997, "A Social Ecology," *Capitalism, Nature, Socialism* 8, 3:3.
Cochrane, R. 1998, "Social Ecology and Reproductive Freedom: A Feminist Perspective," in A. Light. ed., 1998, *Social Ecology After Bookchin*. New York: The Guilford Press.
Cockburn, C. 1985, *The Machinery of Dominance*. London: Pluto.
Cohen, J. and Stewart, M. 1995, *The Collapse of Chaos*. Harmondsworth: Penguin.

Collard, A. and Contrucci, J. 1988, *Rape of the Wild: Man's Violence Against Animals and the Earth*. London: The Women's Press.
Collier, A. 1989, *Scientific Reasoning and Socialist Thought*. Hemel Hempstead: Harvester Wheatsheaf.
Collier, A. 1994, *Critical Realism: An Introduction to Roy Bhaskar's Philosophy*. London: Verso.
Collins, P. 1990, *Black Feminist Thought*. London: Unwin Hyman.
Comninou, M. 1995, "Speech, Pornography and Hunting," in C. J. Adams and J. Donovan. eds, *op cit*.
Connelly, J. and Smith, G. 1999, *Politics and the Environment: From Theory to Practice*. London: Routledge.
Cooper, D. and Palmer, J. 1992, *The Environment in Question*. London: Routledge.
Cooper, D. 1993, "Human Sentiment and the Future of Wildlife," *Environmental Values* 2:335–46.
Cooper, D. and Palmer, J. eds, 1995, *Just Environments: Intergenerational, International and Interspecies Issues*. London: Routledge.
Cooper, D. 1995, "Other Species and Moral Reason," in D. Cooper and J. Palmer. eds, *ibid*.
Corea, G. 1985a, *The Mother Machine*. London: Women's Press.
Corea, G. 1985b, "How the New Reproductive Technologies could be Used to Apply the Brothel Model of Social Control Over Women," *Women's Studies International Forum* 8, 4: 299–305.
Cotgrove, S. 1982, *Catastrophe or Cornucopia: The Environment, Politics, and the Future*. Chichester: John Wiley and Sons.
Cotgrove, S. 1991, "Sociology and the Environment: Cotgrove Replies to Newby," *Network* 51, October, 5.
Coveney, L., Jackson, M., Jeffreys, S., Kaye, L. and Mahoney, P. 1984, *The Sexuality Papers*. London: Hutchinson.
Coward, R. 1983, *Patriarchal Precedents*. London: Routledge & Kegan Paul.
Coward, R. 1984, *Female Desire*. London: Granada.
Coward, R. 1992, *Our Treacherous Hearts: Why Women Let Men Get Their Way*. London: Faber and Faber.
Craig, S. 1994, *Men, Masculinities and the Media*. London: Sage.
Crenshaw, K. 1998, "Demarginalizing the Intersection of Race and Sex: A Black Feminist Critique of Antidiscrimination Doctrine, Feminist Theory and Antiracist Politics," in A. Phillips. ed., *Feminism and Politics*. Oxford: Oxford University Press.
Cronon, W. 1995 "The Trouble with Wilderness, or Getting Back to the Wrong Nature," in W. Cronon. ed., *Uncommon Ground: Toward Reinventing Nature*. New York: W.W. Norton.
Crowley, V. 1989, *Wicca*. London: Thorsons.
Cudworth, E. 1998, *Gender, Nature and Domination*. Unpublished PhD thesis, The University of Leeds.
Cudworth, E. 1999, "The Structure/Agency Debate in Environmental Sociology: Towards a Structural and Realist Approach," *Social Politics Papers*. No. 4. The University of East London.
Cudworth, E. 2003, *Environment and Society*. London: Routledge.
Cuomo, C. J. 1994, "Ecofeminism, Deep Ecology, and Human Population," in K. J. Warren. ed., *Ecological Feminism*. London: Routledge.

Cuomo, C. 1998, *Feminism and Ecological Communities: An Ethic of Flourishing.* London: Routledge.
Dalla Costa, M. 1973, *The Power of Women & the Subversion of the Community.* Bristol: Falling Wall Press.
Daly, M. 1973, *Beyond God the Father.*1986 edition, London: The Women's Press.
Daly, M. 1979, *Gyn/Ecology: The Metaethics of Radical Feminism.* London: The Women's Press.
Daly, M. 1991, *Gyn/Ecology,* New Introduction. London: Women's Press.
Daly, M. 1984, *Pure Lust.* London: Women's Press.
Daly, M. and Caputi, J. 1988, *Websters' First New Intergalactic Wickedary of the English Language.* London: Women's Press.
Daly, M. 1993, *Outercourse: The Be-Dazzling Voyage.* London: The Women's Press.
Daly, M. 2000, *Quintessence: Realizing the Archaic Future.* Boston, MA: Beacon Press.
Davion, V. 1994, "Is Ecofeminism Feminist?," in K. Warren. ed., *Ecological Feminism.* London: Routledge.
Davis, A. 1981, *Women, Race and Class.* London: Women's Press.
Davis, A. 1990, *Women, Culture and Politics.* London: Women's Press.
Davis, K. 1995, "Thinking Like a Chicken: Farm Animals and the Feminine Connection," in C. J. Adams and J. Donovan. eds, *op cit.*
Davis, K. ed., 1997, *Embodied Practices: Feminist Perspectives on the Body.* London: Sage.
d'Eaubonne, F. 1980, "Le feminisme ou la mort," in E. Marks and I. de Courtivron. eds, *New French Feminisms: An Anthology.* Amherst: University of Massachusetts Press.
De Lauretis, T. 1984, *Alice Doesn't: Feminism, Semiotics, Cinema.* Bloomington, IL: Indiana University Press.
De Lauretis, T. ed. 1986, *Feminist Studies/Critical Studies.* London: Macmillan.
De Lauretis, T. 1989, "The Essence of the Triangle or, Taking the Risk of Essentialism Seriously: Feminist Theory in Italy, the U.S. and Britain," *Differences, a Journal of Feminist Cultural Studies* 1, 2:3–37.
De Lauretis, T. 1990, "Upping the Anti (sic) in Feminist Theory," in M. Hirsch and E. Fox-Keller. eds, *Conflicts in Feminism.* London: Routledge.
De Lauretis, T. 1994, *The Practice of Love: Lesbian Sexuality and Perverse Desire.* Bloomington, IL: Indiana University Press.
Deleuze, G. 1988, *Foucault* trans. S. Hand, Minneapolis: University of Minnesota Press.
Deleuze, G. and Guttari, F. 1975, *Kafka: Toward a Minor Literature* trans. D. Polan, Minneapolis: University of Minnesota Press.
Deleuze, G. and Guttari, F. 1987, *A Thousand Plateaus: Capitalism and Schizophrenia* trans. B. Massumi, London: Athlone Press.
Delphy, C. 1984, *Close to Home: A Materialist Analysis of Women's Oppression.* London: Hutchinson.
Delphy, C. 1987, "Protofeminism and Antifeminism," in T. Moi. ed., *French Feminist Thought: A Reader.* Oxford: Basil Blackwell.
Demeritt, D. 2001, "Being Constructive About Nature," in N. Castree and B. Braun. eds, 2001, *Social Nature.* Oxford: Blackwell.
Derrida, J. 1978, "Cogito and the History of Madness," in *Writing and Difference.* London: Routledge and Kegan Paul.
Devall, B. and Sessions, G. 1985, *Deep Ecology: Living as if Nature Mattered.* Layton: Gibbs M. Smith.

Devall, B. 1990, *Simple in Means, Rich in Ends*. London: Greenprint.
Diamond, I. and Quinby, L. eds, 1988, *Feminism and Foucault: Reflections on Resistance*. Boston, MA: Northeastern University Press.
Diamond, I. and Orenstein, G. F. eds, 1990, *Reweaving the World: The Emergence of Ecofeminism*. San Francisco, CA: Sierra Club Books.
Dickens, P. 1992, *Society and Nature: Towards a Green Social Theory*. London: Harvester Wheatsheaf.
Dickens, P. 1996, *Reconstructing Nature: Alienation, Emancipation and the Division of Labour*. London: Routledge.
Dickens, P. 2001, "Linking the Social and Natural Sciences: Is Capital Modifying Human Biology in Its Own Image?," *Sociology* 35,1:93–110.
Dinnerstein, D. 1987, *The Rocking of the Cradle and the Ruling of the World*. London: Women's Press, first published 1976.
Di Stephano, C. 1990, "Dilemmas of Difference: Feminism, Modernity and Postmodernism," in L. Nicholson ed., *Feminism/Postmodernism*. London: Routledge.
Dobash, R. and Emerson, R. 1980, *Violence Against Wives: A Case Against Patriarchy*. London: Open Books.
Dobson, A. ed., 1991, *The Green Reader*. London: Andre Deutsch.
Dobson, A. 1992, *Green Political Thought*. London: Routledge.
Donovan, J. and Adams, C. J. eds, 1996. *Beyond Animal Rights: A Feminist Ethic for the Treatment of Animals*. New York: Continuum.
Doubiago, S. 1989, "Mama Coyote Talks to the Boys," in J. Plant. ed., *Healing the Wounds: The Promise of Ecofeminism*. London: Green Print.
Douglas, M. 1992, *Risk and Blame: Essays in Cultural Theory*. London: Routledge.
Dunayer, J. 1995, "Sexist Words, Speciesist Roots," in C. J. Adams and J. Donnovan. eds, *Animals and Women*. London: Duke University Press.
Dunlap, R. E and Catton, W. R. 1993, "The Development, Current Status and Probable Future of Environmental Sociology: Toward an Ecological Sociology," *The Annals of the International Institute of Sociology* 3:263–84.
Dworkin, A. 1974, *Woman Hating*. New York: E. P. Dutton.
Dworkin, A. 1981, *Pornography*. London: Women's Press.
Dworkin, A. 1983, *Right Wing Women*. London: Women's Press.
Dworkin, A. 1988a, *Intercourse*. London: Arrow Books.
Dworkin, A. 1988b, *Letters from a War Zone*. New York: Secker Warburg.
Eastlea, B. 1981, *Science and Sexual Oppression: Patriarchy's Confrontation with Women and Nature*. London: Weidenfeld and Nicholson.
Eckersley, R. 1989, "Green Politics and the New Class: Selfishness or Virtue?" *Political Studies* 37:205–23.
Eckersley, R. 1992, *Environmentalism and Political Theory: Toward an Ecocentric Approach*. London: University College London Press.
Eckersley, R. 1998, "Divining Evolution and Respecting Evolution," in A. Light. ed., 1998, *Social Ecology After Bookchin*. New York: The Guilford Press.
Eder, K. 1996, *The Social Construction of Nature*. London: Sage.
Edwards, J. *et al*, 1999, *Technologies of Procreation: Kinship in the Age of Assisted Conception*. Second edition. London: Routledge.
Edwards, J. 1999, "Explicit Connections: Ethnographic Enquiry in North-West England" in J. Edwards *et al*. *Technologies of Procreation: Kinship in the Age of Assisted Conception*. Second edition. London: Routledge.

Ehrenreich, B. and English, D. 1973, *Witches, Midwives and Nurses: A History of Women Healers*. London: Writers & Readers.
Eisenstein, Z. R. 1979, "Developing a Theory of Capitalist Patriarchy and Socialist Feminism," in Z. R. Eisenstein. ed., *Capitalist Patriarchy*. New York: Monthly Review Press.
Eisenstein, Z. R. 1981, *The Radical Future of Liberal Feminism*. New York: Longman.
Eisenstein, Z. R. 1984, *Feminism and Sexual Equality*. New York: Monthly Review Press.
Eisler, R. 1990, *The Chalice and the Blade*. London: Unwin.
Elkington, J. and Burke, T. 1987, *The Green Capitalists: Industry's Search for Environmental Excellence*. London: Gollancz.
Elkington, J. and Hailes, J. 1988, *The Green Consumer Guide: From Shampoo to Champagne: High Street Shopping for a Better Environment*. London: Gollancz.
Elliot, R. and Gare, A. eds, *Environmental Philosophy*. Milton Keynes: Open University Press.
Elshtain, J. B. 1987, *Public Man, Private Woman: Women in Social and Political Thought*. Oxford: Martin Robertson.
Elstain, J. B. 1987, *Women and War*. Brighton: Harvester.
English, D., Hollibaugh, A. and Rubin, G. 1987, "Talking Sex: A Conversation on Sex and Feminism," in Feminist Review. ed., *Sexuality: A Reader*. London: Fontana.
Enloe, C. 1983, *Does Khaki Become You? The Militarization of Women's Lives*. London: Pluto.
Escobar, A. 1996, "Constructing Nature: Elements of a Post-Structural Political Ecology," in R. Peet and M. Watts. eds, *Liberation Ecology*. London: Routledge.
Escobar, A. 1999, "After Nature: Steps to an Anti-Essentialist Political Ecology," *Current Anthropology* 40:1–30.
Evans, J. 1995, *Feminist Theory Today: An Introduction to Second Wave Feminism*. London: Sage.
Evans, M. ed., 1982, *The Woman Question: Readings on the Subordination of Women*. London: Fontana.
Evans, M. ed., 1994, *The Woman Question*. Second edition. London: Sage.
Eve, R. A., Horsfall, S. and Lee, M. E. 1997, *Chaos, Complexity and Sociology*. London: Sage.
Faderman, L. 1981, *Surpassing the Love of Men: Romantic Friendship and Love Between Women form the Renaissance to the Present*. London: Junction Books.
Faludi, S. 1992, *Backlash: The Undeclared War Against Women*. London: Chatto & Windus.
Featherstone, M. 1988, "In Pursuit of the Postmodern: An Introduction," *Theory, Culture and Society* 5, 2–3:195–215.
Featherstone, M. ed., 1990, *Global Culture: Nationalism, Globalization, Modernity*. London: Sage.
Felski, R. 1997, "The Doxa of Difference," *Signs* 23,1:1–22.
Feminist Review ed., 1987, *Sexuality: A Reader*. London: Virago.
Ferguson, A. 1989, *Blood at the Root: Motherhood, Sexuality and Male Dominance*. London: Pandora.
Ferguson, K. 1993, *The Man Question*. Berkeley, CA: University of California Press.
Figes, E. 1970, *Patriarchal Attitudes*. London: Macmillan.
Figes, K. 1994, *Because of Her Sex: The Myth of Equality for Women in Britain*. London: Pan Books.

Firestone, S. 1988, *The Dialectic of Sex*. London: Women's Press, first British publication, 1971.
Fisher, E. 1979, *Woman's Creation: Sexual Evolution and the Shaping of Society*. Garden City, NY: Doubleday.
Flax, J. 1987, "Postmodernism and Gender Relations in Feminist Theory," *Signs* 12, 4:621–43.
Flax, J. 1990, *Thinking Fragments: Psychoanalysis, Feminism and Postmodernism in the Contemporary West*. Berkeley, CA: University of California Press.
Flax, J. 1990, "Postmodernism and Gender Relations in Feminist Theory," in L. Nicholson. ed., *Feminism/Postmodernism*. London: Routledge.
Foreman, D. and Haywood, B. eds, 1989, *Ecodefense: A Field Guide to Monkeywrenching*. Tucson: Ned Ludd Books.
Foucault, M. 1971, *Madness and Civilization*. London: Tavistock.
Foucault, M. 1972, *The Archeology of Knowledge* trans. Sheridan, A. London: Tavistock.
Foucault, M. 1976a, Lecture, 7th January 1976, in M. Kelly. ed., 1994, *Critique and Power*. Cambridge: MIT Press.
Foucault, M. 1976b, Lecture, 14th January 1976, in M. Kelly. ed., *ibid*.
Foucault, M. 1979, *Discipline and Punish: The Birth of the Prison* trans. A. Sheridan, Harmondsworth: Penguin.
Foucault, M. 1980, "Body/Power," in C. Gordon. ed., *Michel Foucault: Power/Knowledge*. Brighton: Harvester.
Foucault, M. 1981, *The History of Sexuality, Volume One: An Introduction*. Harmondsworth: Penguin.
Foucault, M. 1985, *The History of Sexuality, Volume Two: The Use of Pleasure*. Harmondsworth: Penguin.
Foucault, M. 1986, *The History of Sexuality, Volume Three: The Care of the Self*. Harmondsworth: Penguin.
Fox, K. M. 1997, "Leisure: Celebration and Resistance in the Ecofeminist Quilt," in K. Warren. ed., *Ecofeminism: Women, Nature, Culture*. Philadelphia, PA: New Society.
Fox, W. 1984, "Deep Ecology: A New Philosophy of Our Time?" *The Ecologist* 14, 5/6.
Fox, W. 1986, *Approaching Deep Ecology*. Tasmania: University of Tasmania.
Fox, W. 1989, "The Deep Ecology – Ecofeminism Debate and its Parallels," *Environmental Ethics* 11:5–25.
Fox, W. 1995, *Towards a Transpersonal Ecology*. Dartington: Green Books.
Frank, A. 1991, "For a Sociology of the Body: An Analytical Review," in M. Featherstone, M. Hepworth and B. Turner. eds, *The Body: Social Process and Cultural Theory*. London: Sage.
Franklin, A. 1999, *Animals and Modern Cultures: A Sociology of Human–Animal Relations in Modernity*. London: Sage.
Franklin, S. 1997, *Embodied Progress: A Cultural Account of Assisted Conception*. London: Routledge.
Fraser, N. and Nicholson, L. 1988, "Social Criticism Without Philosophy: An Encounter between Feminism and Post-Modernism," *Theory, Culture and Society* 5:373–94.
Fraser, N. 1989, *Unruly Practices: Power, Discourse and Gender in Contemporary Social Theory*. Cambridge: Polity.

Fraser, N. and Nicholson, L. 1990, "Social Criticism without Philosophy: An Encounter between Feminism and Postmodernism," in L. Nicholson. ed., *Feminism/Postmodernism*. London: Routledge.
Freer, J. 1983, "Gaea: The Earth as Our Spiritual Heritage," in Caldecott & Leyland 1983. London: The Women's Press.
French, M. 1986, *Beyond Power*. London: Abacus.
Friedan, B. 1965, *The Feminine Mystique*. Harmondsworth: Penguin, first published 1963.
Friedan, B. 1981, *The Second Stage*. London: Michael Joseph.
Frye, M. 1983, *The Politics of Reality: Essays in Feminist Theory*. Trumpensburg, NY: The Crossing Press.
Fukuyama, F. 1992, *The End of History and the Last Man!* London: Hamish Hamilton.
Fuss, D. 1989, *Essentially Speaking: Feminism, Nature and Difference*. London: Routledge.
Gallagher, J. 1987, "Eggs, Embryo's Foetuses: Anxiety and the Law," in M. Stanworth. ed., *Reproductive Technologies*. Cambridge: Polity Press.
Gamman, L. and Marshment, M. eds, 1988, *The Female Gaze*. London: Women's Press.
Gatens, M. 1996, *Imaginary Bodies: Ethics, Power and Corporeality*. London: Routledge.
Gearheart, S. M. 1982, "The Future, If There Is One, Is Female," in P. McAllister. ed., *Reweaving the Web of Life*. Philadelphia, PA: New Society.
Giddens, A. 1971, *Capitalism and Modern Social Theory*. Cambridge: Cambridge University Press.
Giddens, A. 1973. *The Class Structure of Advanced Societies*. London: Hutchinson.
Giddens, A. 1979, *Central Problems in Social Theory: Action, Structure and Contradiction in Social Analysis*. London: Macmillan.
Giddens, A. 1984, *The Constitution of Society*. Cambridge: Polity.
Giddens, A. 1990, *The Consequences of Modernity*. Cambridge: Polity.
Giddens, A. 1991, *Modernity and Self-Identity*. Cambridge: Polity.
Gilligan, C. 1982, *In a Different Voice*. Cambridge, Mass.: Harvard University Press.
Gimbutas, M. 1977, "The First Wave of Eurasian Steppe Pastoralists into Copper Age Europe," *Journal of Indo-European Studies* 5, winter 1977, 277–305.
Gimbutas, M. 1982, *Goddesses and Gods of Old Europe*. Berkeley, CA: University of California Press.
Gimbutas, M. 1990, *The Language of the Goddess*. Berkeley, CA: University of California Press.
Godelier, M. 1984, *The Mental and the Material*. London: Verso.
Goldblatt, D. 1996, *Social Theory and the Environment*. Cambridge: Polity Press.
Goldsmith, E. 1988, *The Great U-Turn: De-Industrializing Society*. Bideford: Green Books.
Goodin, R. E. 1992, *Green Political Theory*. Cambridge: Polity.
Goodman, D. and Redclift, M. 1991, *Refashioning Nature: Food, Ecology and Culture*. London: Routledge.
Gorz, A. 1980, *Ecology as Politics*. London: Pluto.
Gorz, A. 1982, *Farewell to the Working Class: An Essay in Post-Industrial Socialism*. London: Pluto.

Gorz, A. 1985, *Paths to Paradise: On the Liberation from Work*. London: Pluto.
Gorz, A. 1994, *Capitalism, Socialism, Ecology*. London: Verso.
Gottfried, H. 1998, "Beyond Patriarchy? Theorizing Gender and Class," *Sociology* 32, 3:451–68.
Greenwood, S. 1996, "Feminist Witchcraft: A Transformatory Politics," in N. Charles. and F. Hughes-Freeland. eds, *op cit*.
Greer, G. 1971, *The Female Eunuch*. London: Paladin.
Greer, G. 1985, *Sex and Destiny: The Politics of Human Fertility*. London: Pan Books.
Greer, G. 1991, *The Change: Women, Ageing and the Menopause*. London: Hamish Hamilton.
Greer, G. 1999, *The Whole Woman*. London: Doubleday.
Gregory, D. 2001, "(Post)Colonialism and the Production of Nature," in N. Castree and B. Braun. eds, *Social Nature*. Oxford: Blackwell.
Greider, W. 1997, *One World, Ready or Not: The Manic Logic of Global Capitalism*. New York: Simon and Schuster.
Griffin, S. 1982, *Made from This Earth*. London: Women's Press.
Griffin, S. 1983, in Preface to Caldecott, L. and Leyland, S. eds, *Reclaim the Earth*. London: The Women's Press.
Griffin, S. 1984, *Woman and Nature*. London: Women's Press.
Griffin, S. 1988, *Pornography and Silence*. London: Women's Press.
Griffin, S. 1989, "Split Culture," in J. Plant. ed., *Healing the Wounds: The Promise of Ecofeminism*. Philadelphia, PA: New Society Publishers.
Griffin, S. 1994, *A Chorus of Stones: The Private Life of War*. London: Women's Press.
Griffin, S. 1997, "Ecofeminism and Meaning," in K. Warren. ed., *Ecofeminism: Women, Nature, Culture*. Philadelphia, PA: New Society.
Grimshaw, J. 1993, "Practices of Freedom," in C. Ramazanoglu. ed., *Up Against Foucault*. London: Routledge.
Grosz, E. 1994, *Volatile Bodies: Towards a Corporeal Feminism*. London: Allen and Unwin.
Grosz, E. 1995, "Animal Sex. Libido as Desire and Death," in E. Grosz and E. Probyn. eds, *Sexy Bodies: The Strange Carnalities of Feminism*. London: Routledge.
Grosz, E. and Probyn, E. eds, 1995, *Sexy Bodies: The Strange Carnalities of Feminism*. London: Routledge.
Gruen, L. 1994, "Toward an Ecofeminist Moral Epistemology," in K. J. Warren. ed., *Ecological Feminism*. London: Routledge.
Gruen, L. 1997, "Revaluing Nature," in K. Warren. ed., *Ecofeminism: Women, Nature, Culture*. Philadelphia, PA: New Society.
Guha, R. 1989a, "Radical American Environmentalist & Wilderness Protection: A Critique," *Environmental Ethics* 11: 71–83.
Guha, R. 1989b, *The Unquiet Woods: Ecological Change and Peasant Resistance in the Himalayas*. Oxford: Oxford University Press.
Guha, R. 1997, "The Environmentalism of the Poor," in R. Guha and J. Martinez-Alier. eds, *Varieties of Environmentalism*. London: Earthscan.
Gutting, G. 1989. *Michael Foucault's Archaeology of Scientific Reason*. Cambridge: Cambridge University Press.
Habermas, J. 1979, *Communication and the Evolution of Society*. London: Heinemann.

Habermas, J. 1984, *The Theory of Communicative Action Vol.1: Reason and the Rationalization of Society*. Oxford: Polity.
Habermas, J. 1987, *The Theory of Communicative Action Vol.2: A Critique of Functionalist Reason*. Oxford: Polity.
Hajer, M. 1995, *The Politics of Environmental Discourse: Ecological Modernization and the Policy Process*. Oxford: Clarendon Press.
Hakim, C. 1995, "Five Feminist Myths about Women's Employment," *British Journal of Sociology* 46, 3:429–55.
Halberstram, J. 1991, "Automating Gender: Postmodern Feminism in the Age of the Intelligent Machine," *Feminist Studies* 3:439–60.
Halberstram, J. and Livingston, I. 1995, *Posthuman Bodies*. Bloomington, IL: Indiana University Press.
Hall, R. 2002, "When is a Wife not a Wife? Some Observations on the Immigration Experiences of South Asian Women in West Yorkshire," *Contemporary Politics* 8, 1:55–68.
Hall, P. A. and Soskice, D. eds, 2001, *Varieties of Capitalism: The Institutional Foundations of Comparative Advantage*. Oxford: Oxford University Press.
Hanmer, J. 1978, "Violence and the Social Control of Women," in G. Littlejohn, B., Smart, J. Wakefield, and N. Yuval-Davis. eds, *Power and the State*. London: Croom Helm.
Hanmer, J. and Saunders, S. 1984, *Well-Founded Fear: A Community Study of Violence to Women*. London: Hutchinson.
Hanmer, J. 1985, "Transforming Consciousness: Women and the New Reproductive Technologies," in G. Corea and R. Duelli Klein. ed., *Man-Made Women*. London: Hutchinson.
Hanmer, J. 2000, "The Common Market of Violence," in M. Rossilli. ed., *Gender Policies in the European Union*. New York: Peter Lang.
Hanmer, J. and Maynard, M. eds, 1987, *Women, Violence and Social Control*. London: Macmillan.
Hanmer, J. 1989, "Women and Policing in Britain," in J., Hanmer, J. Radford and E. A. Stanko. eds, *Women, Policing and Male Violence*. London: Routledge.
Hannigan, J. 1995, *Environmental Sociology: A Social Constructionist Perspective*. London: Routledge.
Haraway, D. 1976, *Crystals, Fabrics and Fields: Metaphors of Organicisim in Twentieth Century Developmental Biology*. New Haven, CT: Yale University Press.
Haraway, D. 1988, "Situated Knowledges: The Science Question in Feminism and the Privilege of Partial Perspective," *Feminist Studies* 14, 3:575–99.
Haraway, D. 1989, *Primate Visions: Gender, Race and Nature in the World of Modern Science*. London: Routledge.
Haraway, D. 1991, *Simians, Cyborgs and Women: The Reinvention of Nature*. London: Free Association Press.
Haraway, D. 1992, "The Promises of Monsters: A Regenerative Politics for Inappropriate/d Others," in L. Grossberg, C. Nelson and PA. Treichler. eds, *Cultural Studies*. New York: Routledge.
Haraway, D. 1994, "A Game of Cat's Cradle: Science Studies, Feminist Theory, Cultural Studies," *Configurations* 1:59–71.
Haraway, D. 1997, *Modest_Witness@Second_Millennium. FemaleMan_Meets_OncoMouse*. London: Routledge.
Haraway, D. 2003, *The Companion Species Manifesto: Dogs, People and Significant Otherness*. Chicago: Prickly Paradigm Press.

Hardin, G. 1977, "The Tragedy of the Commons," in G. Hardin and J. Baden. eds, *Managing the Commons*. San Francisco: W. H. Freeman and Co.
Harding, S. 1986, *The Science Question in Feminism*. Ithaca: Cornell University Press.
Harding, S. 1987, *Feminism and Methodology*. Milton Keynes: Open University Press.
Harding, S. 1990, "Feminism, Science and the Anti-Enlightenment Critiques," in L. Nicholson. ed., *Feminism/Postmodernism*. London: Routledge.
Harding, S. 1991, *Whose Science? Whose Knowledge?*. Milton Keynes: Open University Press.
Harding, S. 1993, "Rethinking Standpoint Epistemology: What is Strong Objectivity?" in L. Alcoff and E. Potter. eds, *Feminist Epistemologies*. London: Routledge.
Harding, S. 1994, "Feminism and Theories of Scientific Knowledge," in M. Evans. ed., *The Woman Question*. Second edition. London: Sage.
Hartmann, B. 1987, *Reproductive Rights and Wrongs: The Global Politics of Population Control and Contraceptive Choice*. New York: Harper & Row.
Hartmann, H. I. 1979, "Capitalism, Patriarchy, and Job Segregation by Sex," in Z. Eisenstein. ed., *Capitalist Patriarchy*. New York: Monthly Review Press.
Hartmann, H. I. 1981, "The Unhappy Marriage of Marxism and Feminism: Towards a More Progressive Union," in L. Sargent. ed., *Women and Revolution*. London: Pluto Press.
Hartsock, N. 1987, "The Feminist Standpoint: Developing the Ground for a Specifically Feminist Historical Materialism," in S. Harding. ed., *Feminism and Methodology*. Milton Keynes: Open University Press.
Hartsock, N. 1990, "Foucault on Power: A Theory for Women?" in L. Nicholson. ed., *Feminism/Postmodernism*. London: Routledge.
Harvey, D. 1989, *The Condition of Postmodernity*. Oxford: Basil Blackwell.
Harvey, D. 1996, *Justice, Nature, and the Geography of Difference*. Oxford: Blackwell.
Hayles, N. K. 1990, *Chaos Bound: Orderly Disorder in Contemporary Literature and Science*. Ithaca: Cornell University Press.
Hayles, N. K. ed., 1991, *Chaos and Order: Complex Dynamics in Literature and Science*. Chicago: University of Chicago Press.
Hayles, N. K. 1999, *How We Became Posthuman: Virtual Bodies in Cybernetics, Literature and Informatics*. Chicago: The University of Chicago Press.
Hayward, T. 1990, "Ecosocialism – Utopian and scientific," *Radical Philosophy* 56, Autumn:2–14.
Hearn, J. 1987, *The Gender of Oppression*. Brighton: Wheatsheaf.
Hekman, S. 1990, *Gender and Knowledge: Elements of a Postmodern Feminism*. Cambridge: Polity.
Held, D. 1995, *Democracy and the Global Order: From the Modern State to Cosmopolitan Governance*. Cambridge: Polity Press.
Held, D., McGrew, A. G., Goldblatt, D. and Perraton, J. 1999, *Global Transformations*. Cambridge: Polity Press.
Henderson, H. 1983, "The Warp and the Weft: The Coming Synthesis of Eco-Philosophy and Eco-Feminism," in Caldecott and Leyland 1983 *op cit*.
Hennessey, R. 1993, *Materialist Feminism and the Politics of Discourse*. London: Routledge.
Hirsch, E. 1999, "Negotiated Limits: Interviews in South-East England," in Edwards, J., Franklin, S., Hirsch, E., Price, F. and Strathern, M. eds, *Technologies*

of Procreation: Kinship in the Age of Assisted Conception. Second edition. London: Routledge.
Hirst, P. and Thompson, G. 1996, *Globalization in Question: The International Economy and the Possibilities of Governance*. Cambridge: Polity Press.
Hoff, J. 1996, "The Pernicious Effect of Post-Structuralism on Women's History," in D. Bell and R. Klein. eds, *Radically Speaking*. London: Zed Books.
hooks, b. 1982, *Ain't I a Woman?* London: Pluto Press.
hooks, b. 1984, *Feminist Theory: From Margin to Center*. Boston, MA: South End Press.
hooks, b. 1991, *Yearning: Race, Gender and Cultural Politics*. London: Turnaround.
hooks, b. 1992, *Black Looks: Race and Representation*. London: Turnaround.
Hubbard, R. and Lowe, M. 1983, *Woman's Nature*. New York: Pergammon.
Hubbard, R. 1990, *The Politics of Women's Biology*. New Brunswick, NJ.: Rutgers University Press.
Ingold, T. 1986, *The Appropriation of Nature*. Manchester: Manchester University Press.
Ingold, T. 1994, "From Trust to Dominion: An Alternative History of Human-Animal Relations," in A. Manning and J. Serpell. eds, *Animals and Human Society*. London: Routledge.
Ingold, T. 2000, *The Perception of the Environment: Essays on Livelihood, Dwelling and Skill*. London: Routledge.
Irigaray, L. 1974, *Speculum of the Otherb Woman* trans. 1985 Gillian Gill, Ithaca: Cornell University Press.
Irigaray, L. 1977, *This Sex Which is not One* 1985 trans. Catherine Porter, Ithaca: Cornell University Press.
Irigaray, L. 1996, *I Love to You*. London: Routledge.
Irvine, S. and Ponton, A. 1988, *A Green Manifesto: Policies for a Green Future*. London: Macdonald Optima.
Irvine, S. 1989, *Beyond Green Consumerism*. London: Friends of the Earth.
Irwin, A. 2001, *Sociology & the Environment*. Cambridge: Polity.
Itzin, C. 1992, "Pornography and the Social Construction of Sexual Inequality," in C. Itzin. ed., *Pornography: Women, Violence and Civil Liberties*. Oxford: Oxford University Press.
Jackson, C. 1994, "Gender Analysis and Environmentalisms" in M. Redclift and T. Benton. eds, *op cit*.
Jackson, S. 1996, "Heterosexuality as a Problem for Feminist Theory," in L. Adkins and V. Merchant. eds, *Sexualizing the Social: Power and the Organization of Sexuality*. Explorations in Sociology 47. London: Macmillan.
Jaggar, A. 1983, *Feminist Politics and Human Nature*. Brighton: Harvester.
Jain, S. 1991, "Standing Up for Trees: Women's Role in the Chipko Movement," in Sontheimer, S. ed., *Women and the Environment: A Reader*. London: Earthscan.
Jameson, F. 1972, *The Prison House of Language*. Princeton, NJ.: Princeton University Press.
Jameson, F. 1984, "Postmodernism or the Cultural Logic of Late Capitalism," *New Left Review* 146: 53–93.
Jamison, A., Eyerman, R. and Cramer, J. 1990, *The Making of the New Environmental Consciousness: A Comparative Study of the Environmental Movements in Sweden, Denmark and the Netherlands*. Edinburgh: Edinburgh University Press.

Jayawardena, K. 1986, *Feminism and Nationalism in the Third World*. London: Zed Press.

Jeffreys, S. 1985, *The Spinster and Her Enemies: Feminism and Sexuality 1880–1930*. London: Pandora.

Jeffreys, S. 1990, *Anti-Climax: A Feminist Perspective on the Sexual Revolution*. London: Women's Press.

Jeffreys, S. 1994, *The Lesbian Heresy: A Feminist Perspective on the Lesbian Sexual Revolution*. London: The Women's Press.

Jeffreys, S. 1996, "The Return to Gender: Post-Modernism and Lesbainandgay Theory," in D. Bell and R. Klein. eds, *op cit*.

Johnson, A. 1995, "Barriers to Fair Treatment of Non-Human Life," in D. Cooper and J. Palmer. eds, *op cit*.

Johnson, D. K. and Johnson, K. R. 1994, "The Limits of Partiality: Ecofeminism, Animal Rights, and Ecological Concern," in K. J. Warren. ed., *Ecological Feminism*. London: Routledge.

Jones, M. and Wangari, M. 1983, "Greening the Desert: Women of Kenya Reclaim the Land," in L. Caldecott and S. Leyland. eds, *Reclaim the Earth*. London: The Women's Press.

Joseph, G. 1981, "The Incompatible Ménage a trois: Marxism, Feminism and Racism," in L. Sargent. ed., *Women and Revolution*. London: Pluto Press.

Kappeler, S. 1986, *The Pornography of Representation*. Cambridge: Polity.

Kappeler, S. 1992, "Pornography: The Representation of Power," in C. Itzin. ed., *Pornography: Women, Violence and Civil Liberties*. Oxford: Oxford University Press.

Kappeler, S. 1995, "Speciesism, Racism, Nationalism … or the Power of Scientific Subjectivity," in C. J. Adams and J. Donovan. eds, *op cit*.

Katherine, A. 1998, "A Too-Early Morning: A Reading of Audre Lourde's Open Letter to Mary Daly," in M. Frye and Hoagland. eds. *Feminist Interpretations of Mary Daly*. Philadelphia, PA: Pennsylvania State University Press.

Kauffman, S. 1993, *The Origins of Order: Self-Organization and Selection in Evolution*. Oxford: Oxford University Press.

Kauffman, S. 1995, *At Home in the Universe: The Search for Laws of Self-Organization and Complexity*. Oxford: Oxford University Press.

Kaw, E. 2003, "Medicalization of Racial Features: Asian-American Women and Cosmetic Surgery," in R. Weitz. ed., *The Politics of Women's Bodies: Sexuality, Appearance and Behavior*. Second edition. Oxford: Oxford University Press.

Keil, L. D. and Elliot, E. eds, *Chaos Theory in the Social Sciences*. Ann Arbor: University of Michigan Press.

Keller, E. F. 1985, *Reflections on Gender and Science*. New Haven, CT: Yale University Press.

Keller, E. F. 1992, *Secrets of Life, Secrets of Death*. London: Routledge.

Keller, E. F. 1995, *Refiguring Life: Metaphors of Twentieth-Century Biology*. New York: University of Colombia Press.

Keller, E. F. 2000, *Century of the Gene*. Cambridge, Mass.: Harvard University Press.

Kellner, D. 1988, "Postmodernism as Social Theory: Some Challenges and Problems," *Theory, Culture and Society*, 5:239–69.

Kelly, L. 1988, *Surviving Sexual Violence*. Cambridge: Polity Press.

Kelly, L. 1999, "A Policy of Neglect or a Neglect of Policy," in S. Walby. ed., *New Agendas for Women*. Basingstoke: Macmillan.

Kelly, M. ed. 1994, *Critique and Power: Recasting the Foucault/Habermas Debate*. Cambridge: MIT Press.
Kelly, P. 1984, *Fighting for Hope*. London: Chatto and Windus.
Kelly, P. 1994, "Women and Power," in K. Warren. ed., *Ecofeminism: Women, Nature, Culture*. Philadelphia, PA: New Society, 1997.
Kemp, P. and Wall, D. 1989, *A Green Manifesto for the 1990's*. London: Greenprint.
Kheel, M. 1995, "License to Kill: An Eco-feminist Critique of Hunters' Discourse," in C. J. Adams and J. Donovan. eds. *ibid*.
Kimbrell, A. 1993, *The Human Body Shop: The Engineering and Marketing of Life*. London: HarperCollins.
Kimmel, M. ed., 1987, *Changing Men*. London: Sage.
King, D. 1988, "Multiple Jeopardy, Multiple Consciousness: The Context of Black Feminist Ideology," *Signs* 14, 1: 42–72.
King, Y. 1983, "The Eco-Feminist Imperative," in Caldecott and Leyland, 1983 *op cit*.
King, Y. 1987, "The Ecology of Feminism and the Feminism of Ecology," in J. Plant. ed., *Healing the Wounds: The Promise of Ecofeminism*. Philadelphia, PA: New Society Publishers.
King, Y. 1990, "Healing the Wounds: Feminism, Ecology and Nature/Culture Dualism," in I. Diamond, and G. Orenstein. eds, *Reweaving the World*. San Francisco, CA: Sierra Club Books.
Kitsuse, J. I. and Spector, M. 1981, "The Labeling of Social Problems," in E. Rubington and M. S. Weinberg. eds, *The Study of Social Problems: Five Perspectives*. New York: Oxford University Press.
Klein, R. ed., 1986, *Infertility: Women Speak Out About Their Experiences of Reproductive Medicine*. London: Pandora.
Klein, R. 1996, "Dead Bodies Floating in Cyberspace: Post-modernism and the Dismemberment of Women," in D. Bell and R. Klein. eds, *Radically Speaking*. London: Zed Books.
Kolakowski, L. 1987, *Main Currents of Marxism: 1. The Founders*. Oxford: Oxford University Press.
Kovel, J. 1998, "Negating Bookchin" in A. Light. ed., 1998, *Social Ecology After Bookchin*. New York: The Guilford Press.
Kropotkin, P. 1955, *Mutual Aid*. New York: Extending Horizons Books.
Lappe, F. M. and Collins, J. 1982, *Food First: Beyond the Myth of Scarcity*. London: Abacus.
Laqueur, T. 1990, *Making Sex: Body and Gender from the Greeks and Freud*. Cambridge, Mass.: Harvard University Press.
Lash, S. 1990, *Sociology of Post-Modernism*. London: Routledge.
Lash, S. 2000, "Risk Culture," in U. Beck, 2000, "Risk Society Revisited," in B. Adam, U. Beck, and J. van Loon, J. *The Risk Society and Beyond: Critical Issues for Social Theory*. London: Sage.
Lash, S. and Urry, J. 1987, *The End of Organized Capitalism*. Cambridge: Polity.
Lash, S. and Urry, J. 1994, *Economies of Signs and Space*. London: Sage.
Lash, S., Szersynski, B. and Wynne, B. eds, 1996, *Risk, Environment and Modernity*. London: Sage.
Lash, S., Szerszynski, B. and Wynne, B. 1996, "Ecology, Realism and the Social Sciences," in S. Lash *et al. op cit*.
Latouche, S. 1991, *In the Wake of the Affluent Society*. London: Zed.
Latour, B. 1987, *Science in Action*. Milton Keynes: Open University Press.

Latour, B. 1993, *We Have Never Been Modern*. Hemel Hempstead: Harvester Wheatsheaf.
Latour, B. 1998, "To Modernise or Ecologise?" in B. Braun and N. Castree. eds, *Remaking Reality: Nature at the Millennium*. London: Routledge.
Latour, B. 1999, *Pandora's Hope: Essays on the Reality of Science Studies*. London: Harvard University Press.
Layder, D. 1996, "Review Essay: Contemporary Social Theory," *Sociology* 30, 3:601–08.
Leakey, R. E. and Lewin, R. 1979, *Origins*. London: MacDonald and Jones.
Lee, J. 2003, "Menarche and the (Hetero)sexualization of the Female Body," in R. Weitz. ed., *The Politics of Women's Bodies: Sexuality, Appearance and Behavior*. Second edition. Oxford: Oxford University Press.
Leeds Revolutionary Feminist Group 1981, "Political Lesbianism: The Case Against Heterosexuality," in Onlywomen. ed., *Love Thy Enemy? The Debate Between Heterosexual Feminism and Political Lesbianism*. London: Onlywomen Press.
Lefebvre, H. 1991, *The Production of Space*. Oxford: Blackwell.
Leff, E. 1996, "Marxism and the Environmental Question: From the Critical Theory of Production to an Environmental Rationality for Sustainable Development," in T. Benton. ed., *The Greening of Marxism*. New York: The Guilford Press.
Leopold, A. 1949, *A Sand Country Almanac and Sketches Here and There*. Oxford: Oxford University Press.
Lerner, G. 1986, *The Creation of Patriarchy*. Oxford: Oxford University Press.
Levitas, R. 2000, "Discourses of Risk and Utopia," in U. Beck, 2000 "Risk Society Revisited" in B. Adam, U. Beck, and J. van Loon, J. *The Risk Society and Beyond: Critical Issues for Social Theory*. London: Sage.
Leyland, S. 1983, "Feminism and Ecology: Theoretical Considerations," in L. Caldecott and S. Leyland. eds, *Reclaim the Earth*. London: The Women's Press.
Lienert, T. 1996, "On Who is Calling Radical Feminist 'Cultural Feminists' and Other Historical Sleights of Hand," in D. Bell and R. Klein. eds, *Radically Speaking*. London: Zed Books.
Light, A. 1998, "Bookchin as/and Social Ecology," in A. Light. ed., 1998, *Social Ecology After Bookchin*. New York: The Guilford Press.
Light, A. ed., 1998, *Social Ecology After Bookchin*. New York: The Guilford Press.
Lockwood, D. 1964, "Social Integration and System Integration," in G. K. Zollschan and W. Hirsch. eds, *Explorations in Social Change*. London: Routledge.
López, J. and Scott, J. 2000, *Social Structure*. Buckingham: Open University Press.
Lorber, J. 1994, *Paradoxes of Gender*. London: Yale University Press.
Lorber, J. 1993, "Believing is Seeing: Biology as Ideology," *Gender and Society* 7, 4:568–81.
Lorde, A. 1981, "An Open Letter to Mary Daly," in C. Moraga and G. Azadula. eds, *This Bridge Called My Back: Writings by Radical Women of Color*. Waterdown: Persephone Press.
Lorde, A. 1994, "Age, Race, Class and Sex: Women Redefining Difference," in M. Evans. ed., *The Woman Question*. Second edition. London: Sage.
Lovejoy, T. E. 1986, "Species Leave the Ark: One by One," in B. G. Norton. ed., *The Preservation of Species: The Value of Biological Diversity*. Princeton NJ.: Princeton University Press.

Lovelock, J. 1979, *Gaia: A New Look at Life on Earth*. Oxford: Oxford University Press.
Lovelock, J. 2000, *The Ages of Gaia: A Biography of Our Living Earth*. Second Edition. Oxford: Oxford University Press.
Lovibond, S. 1989, "Feminism and Postmodernism," *New Left Review* 178, 2:5–28.
Lowe, M. and Hubbard, R. 1983, *Woman's Nature*. London: Tavistock.
Lowe, P. and Rudig, W. 1986, "Review Article: Political Ecology and the Social Sciences," *British Journal of Political Science* 16:513–50.
Luhmann, N. 1990, "The Autopoiesis of Social Systems," in N. Luhman. *Essays on Self Reference*. New York: Colombia University Press.
Luhmann, N. 1993, *Risk: A Social Theory*. New York: Stanford University Press.
Luhmann, N. 1995, *Social Systems*. Cambridge: Cambridge University Press. (first published in 1984).
Luhmann, N. 1998, *Observations on Modernity*. Cambridge: Cambridge University Press.
Luke, B. 1995, "Taming Ourselves or Going Feral? Toward a Non-Patriarchal Metaethic of Animal Liberation," in C. J. Adams and J. Donnovan. eds. *op cit.*
Lukes, S. 1974, *Power: A Radical View*. London: Macmillan.
Lupton, D. 1996, *Food, the Body and the Self*. London: Sage.
Lury, C. 1990, *Cultural Industries: A Political Economy*. London: Routledge.
Lykke, N. 1996, "Between Monsters, Goddesses and Cyborgs: Feminist Confrontations with Science," in N. Lykke and R. Braidotti. eds, *Between Monsters, Goddesses and Cyborgs: Feminist Confrontations with Science, Medicine and Cyberspace*. London: Zed Books.
Lyotard, J. F. 1984, *The Postmodern Condition*. Manchester: Manchester University Press.
Lyotard, J. F. 1988, *The Differend: Phrases in Dispute*. Minneapolis: University of Minnesota Press.
Macauley, D. 1998, "Evolution and Revolution: The Ecological Anarchism of Kropotkin and Bookchin," in A. Light. ed., 1998 *Social Ecology After Bookchin*. New York: The Guilford Press.
MacKinnon, C. 1982 "Feminism, Marxism, Method and the State," *Signs* 7, 3: 515–44.
MacKinnon, C. 1987, *Feminism Unmodified: Discourses on Life and Law*. Cambridge, Mass.: University of Harvard Press.
MacKinnon, C. 1989, *Toward a Feminist Theory of the State*. Cambridge Mass.: University of Harvard Press.
MacKinnon, C. 1994, *Only Words*. London: HarperCollins.
MacKinnon, C. 1996, "From Practice to Theory, or What is a White Woman Anyway?" in D. Bell and R. Klein. eds, *Radically Speaking*. London: Zed Books.
Macnaghten, P. and Urry, J. 1995, "Towards a Sociology of Nature," *Sociology* 29, 2:203–20.
Macnaghten, P. and Urry, J. 1998, *Contested Natures*. London: Sage.
Maguire, J. 1996, "The Tears Inside the Stone: Reflections on the Ecology of Fear," in S. Lash *et al. op cit.*
Mahony, P. and Zmroczek, C. 1996 "Working-Class Radical Feminism," in Bell and Klein. eds, *op cit.*
Malos, E. ed., 1980, *The Politics of Housework*. London: Allison and Busby.
Manes, C. 1990, *Green Rag*. Boston, MA: Little Brown.

Marcuse, H. 1972, *Counter-Revolution and Revolt*. Boston, MA: Beacon Press.
Margulis, L. 1995, "Gaia is a Tough Bitch," in J. Brockman. ed., *The Third Culture*. New York: Simon and Schuster.
Margulis, L. and Sagan, D. 1986, *Microcosmos*. New York: Summit.
Margulis, L. and Sagan, D. 1995, *What is Life?* New York: Simon and Schuster.
Marks, E. and de Courtivron, I. eds, 1981, *New French Feminisms*. Brighton: Harvester.
Marshall, D. 1993, *Demanding the Impossible: A History of Anarchism*. London: Macmillan.
Martell, L. 1994, *Ecology and Society: An Introduction*. Cambridge: Polity.
Martin, E. 1996, "Citadels, Rhizomes and String Figures," in A. Aronowitz, B. Martinsons and M. Menser. eds, *Technoscience and Cyberculture*. London: Routledge.
Marx, K. 1976, *Capital*. Harmondsworth: Penguin.
Marx, K. and Engels, F. 1983, *Manifesto of the Communist Party*. Harmondsworth: Penguin.
Maturana, H. and Varela, F. J. 1980, *Autiopoisis and Cognition: The Realization of the Living*. London: Kulwer Academic Publishers.
Maturana, H. and Varela, F. J. 1987, *The Tree of Knowledge: The Biological Roots of Human Understanding*. Boston: Shambhala.
Maynard, M. 1994, " 'Race', Gender and the Concept of Difference in Feminist Thought," in H. Afshar and M. Maynard. eds, *Dynamics of "Race" and Gender*. London: Taylor and Francis.
McAllister, P. ed., 1982, *Reweaving the Web of Life: Feminism and Non-violence*. Philadelphia, PA: New Society Publishers.
McGrew, A. G. ed., 1997, *The Transformation of Democracy? Globalization and Territorial Democracy*. Cambridge: Polity Press.
McIntosh, M. 1978, "The State and the Oppression of Women," in M. Evans. ed., 1982, *The Woman Question*. London: Fontana.
McIntosh, M. 1992, "Liberalism and the Contradictions of Sexual Politics," in L. Segal and M. McIntosh. eds. *Sex Exposed*. London: Virago.
McLellan, G. 1995, "Feminism, Epistemology and Postmodernism: Reflections on Current Ambivalence," *Sociology* 29, 2:391–409.
McNay, L. 1992, *Foucault and Feminism: Power, Gender and the Self*. Cambridge: Polity.
McNay, L. 1994, *Foucault: A Critical Introduction*. Cambridge: Polity.
McNeil, M, Varcoe, I. and Yearley, S. eds. 1990 *The New Reproductive Technologies*. London: Macmillan.
McNeil, M. 1993, "Dancing with Foucault: Feminism and Power Knowledge," in C. Ramazanoglu. ed., *Up Against Foucault*. London: Routledge.
Meadows, D., Randers, H. J. and Behrens, W. W. 1972, *The Limits to Growth*. New York: Universe Books.
Meadows, D. H., Meadows, D. L. and Randers, J. 1992, *Beyond the Limits: Global Collapse or a Sustainable Future: Sequel to the Limits to Growth* London: Earthscan.
Mellaart, J. 1965, *Earliest Civilizations of the Near East*. London: Thames & Hudson.
Mellaart, J. 1967, *Catal Huyuk*. London: Thames & Hudson.
Mellaart, J. 1975, *The Neolithic of the Near East*. New York: Scribner.
Mellor, M. 1992, *Breaking the Boundaries: Towards a Feminist Green Socialism*. London: Virago Press.

Mellor, M. 1997, *Feminism and Ecology*. Cambridge: Polity.
Melucci, A. 1989, *Nomads of the Present: Social Movements and Individual Needs in Contemporary Society*. London: Radius.
Merchant, C. 1980, *The Death of Nature: Women, Ecology and the Scientific Revolution*. San Francisco, CA: Harper & Row.
Merchant, C. 1985, *Ecological Revolutions*. New York: Harper & Row.
Merchant, C. 1990, "Ecofeminism and Feminist Theory," in I. Diamond, and G. F. Orenstein. eds. *Reweaving the World: The Emergence of Ecofeminism*. San Francisco: Sierra Club Books.
Merchant, C. 1992, *Radical Ecology*. London: Routledge.
Middleton, N, O'Keefe, P. and Moyo, S., 1993, *Tears of the Crocodile: From Rio to Reality in the Developing World*. London: Pluto.
Midgley, M. 1983, *Animals and Why They Matter: A Journey Around the Species Barrier*. Harmondsworth: Penguin.
Midgley, M. 1994, "Bridge-Building at Last," in A. Manning and J. Serpell. eds. *Animals and Human Society*. London: Routledge.
Midgley, M. 1996, *Utopias, Dolphins and Computers: Problems of Philosophical Plumbing*. London: Routledge.
Mies, M. 1986, *Patriarchy and Accumulation on a World Scale*. London: Zed Press.
Mies, M., Bennholdt-Thompson, V. and von Werlhof, C. 1988a, *Women: The Last Colony*. London: Zed Books.
Mies, M. 1988b, "From the Individual to the Dividual: In the Supermarket of Reproductive Alternatives," *Reproductive and Genetic Engineering* 1, 3:225–37.
Mies, M. 1993, "Liberating the Consumer," in M. Mies and V. Shiva, 1993, *Ecofeminism*. London: Zed Books.
Mies, M. 1993, "The Need for a New Vision: the Subsistence Perspective," in M. Mies and V. Shiva, 1993, *Ecofeminism*. London: Zed Books.
Mies, M. and Shiva, V. 1993 *Ecofeminism*. London: Zed Books.
Millett, K. 1985, *Sexual Politics*. London: Virago first published in Britain, 1970.
Milton, K. 2002, *Loving Nature: Towards an Ecology of Emotion*. London: Routledge.
Mirza, H. 1997, "Introduction: Mapping a Genealogy of Black British Feminism," in H. Mirza. ed., *Black British Feminism: A Reader*. London: Routledge.
Mirza, H. ed., *Black British Feminism: A Reader*. London: Routledge.
Mitchell, J. 1971, *Woman's Estate*. Harmondsworth: Penguin.
Mitchell, J. 1974, *Psychoanalysis and Feminism*. London: Allen Lane.
Mitchell, J. and Oakley, A. eds, 1976, *The Rights and Wrongs of Women*. Harmondsworth: Penguin.
Mitchell, J. 1977, *Women, the Longest Revolution*. Harmondsworth: Penguin.
Mitford, J. 1992, *The American Way of Birth*. London: Victor Gollancz.
Moore, H. L. 1994, *A Passion for Difference: Essays in Anthropology and Gender*. Cambridge: Polity.
Morgan, E. 1972, *The Descent of Woman*. New York: Stein & Day.
Morgan, K. P. 1991, "Women and the Knife: Cosmetic Surgery and the Colonization of Women's Bodies," *Hypatia* 6, 3:25–53.
Mouzelis, N. 1991, *Back to Sociological Theory: The Construction of Social Orders*. London: Macmillan.
Mouzelis, N. 1995, *Sociological Theory: What Went Wrong?*. London: Macmillan.
Mouzelis, N. 1997, "Social and System Integration: Lockwood, Habermas and Giddens" *Sociology* 31, 1:111–19.

Murcott, A. 1988, "On the Altered Appetites of Pregnancy: Conceptions of Food, Body and Person," *Sociological Review* 36, 4:733–64.
Murdock, J. 1997, "Inhuman/Nonhuman/Human," *Environment and Planning D: Society and Space* 15:731–56.
Murdock, J. 2001, "Ecologising Sociology: Actor-Network Theory, Co-constructionism and the Problem of Human Exemptionalism," *Sociology* 35, 1:111–33.
Murphy, R. 1994a, "The Sociological Construction of Science Without Nature," *Sociology* 28,4:957–74.
Murphy, R. 1994b, *Rationality and Nature*. Boulder, CO: Westview Press.
Naess, A. 1973, "The Shallow and the Deep, Long-Range Ecology Movement: A Summary," *Inquiry* 16:95–100.
Naess, A. 1979, "Self-Realization in Mixed Communities of Humans, Bears, Sheep and Wolves," *Inquiry* 16:95–100.
Naess, A. 1984, "Intuition, Intrinsic Value and Deep Ecology," *The Ecologist* 14:5–6.
Naess, A. 1985, "Identification as a Source of Deep Ecology Attitudes" in M. Tobias. ed., *Deep Ecology*. San Diego: Avant Books.
Naess, A. 1989, *Ecology, Community and Lifestyle: Outline of an Ecosophy*. Cambridge: Cambridge University Press.
Naess, A. 1990, "Deep Ecology" in A. Dobson. ed., *The Green Reader*. London: Andre Deutsch.
Naess, A. 1994, "Ecosophy T: Deep versus Shallow Ecology" in L. Pojman. ed. *Environmental Ethics: Readings in Theory and Application*. Boston: Jones and Bartlett.
Naess, A., Fleming, P. Macy, J. and Steed, J. 1988. *Thinking Like a Mountain*. Santa Cruz: New Society Publishers.
Nash, C. 2000, "Performativity in Practice: Some Recent Work in Cultural Geography," *Progress in Human Geography* 24, 4:653–64.
Neal, B. and Smart, C. 1997, "Experiments with Parenthood?," *Sociology* 31, 2:201–21.
Nelson, L. 1999, "Bodies (and Spaces) do Matter: The Limits of Performativity," *Gender, Place and Culture* 6, 4:331–53.
New, C. 1998, "Realism, Deconstruction and the Feminist Standpoint," *Journal for the Theory of Social Behaviour* 28, 4:349–73.
Newby, H. 1991, "One World, Two Cultures: Sociology and the Environment," BSA Bulletin, *Network* 50, May:1–8.
Nicholson, L. J. ed. 1990, *Feminism/Postmodernism* London: Routledge.
Nicholson, L. 1992, "On the Postmodern Barricades: Feminism, Politics and Social Theory," in S. Seidman and D. G. Wagner. eds, *Postmodernism and Social Theory*, Oxford: Basil Blackwell.
Oakley, A. 1972, *Sex, Gender and Society*. London: Maurice Temple Smith.
Oakley, A. 1976, "Wisewoman and Medicine Men: Changes in the Management of Childbirth," in J. Mitchell and A. Oakley. eds, *The Rights and Wrongs of Women*. Harmondsworth: Penguin.
Oakley, A. 1976, *Housewife*. Harmondsworth: Penguin.
Oakley, A. 1984, *The Captured Womb*. Oxford: Basil Blackwell.
Oakley, A. 1987, "From Walking Wombs to Test Tube Babies," in M. Stanworth, ed., *Reproductive Technologies*. Cambridge: Polity Press.
Oakley, A. 2002, *Gender on Planet Earth*. Cambridge: Polity.

O'Brien, M. 1981, *The Politics of Reproduction*. London: Routledge and Kegan Paul.
O'Connor, J. 1989, "Political Economy and the Ecology of Socialism and Capitalism," *Capitalism, Nature, Socialism* 1, 1:11–38.
O'Connor, J. 1994, "On the Misadventures of Capitalist Nature," in J. Connor. ed., *Is Capitalism Sustainable?* London: Guilford Press.
Offe, C. 1985, *Disorganized Capitalism: Contemporary Transformations of Work and Politics*. Cambridge: Cambridge University Press.
Ohmae, K. 1990, *The Borderless World*. London: Collins.
Ohmae, K. 1995, *The End of the Nation State*. New York: Free Press.
Okin, S. M. 1990, *Gender, Justice and the Family*. New York: Basic Books.
O'Neill, J. 1993, *Ecology, Policy and Politics*. London: Routledge.
O'Riordan, T. 1981, *Environmentalism*. London: Pion.
O'Sullivan, S. 1987, "Passionate Beginnings: Ideological Politics 1969–72," in Feminist Review. ed., *Sexuality: A Reader*. London: Virago.
Paglia, C. 1991, *Sexual Personae*. New York: Vintage.
Pahl, J. ed. 1985, *Private Violence and Public Policy*. London: Routledge.
Palmer, J. and Cooper, D. E. 1998, *Spirit of the Environment: Religion, Value and Environmental Concern*. London: Taylor and Francis.
Palmer, J. ed., 2001, *50 Key Thinkers on the Environment*. London: Routledge.
Parsons, T. 1960, *Structure and Process in Modern Societies*. Glencoe: The Free Press.
Passmore, J. 1980, *Man's Responsibility for Nature*. Second edition. London: Duckworth.
Pateman, C. 1988, *The Sexual Contract*. Cambridge: Polity.
Pateman, C. 1989, *The Disorder of Women*. Cambridge: Polity.
Pearce, F. 1991, *Green Warriors*. London: Bodley Head.
Peet, R. 1985 "The Social Origins & Environmental Determinism," *Annals & the Association & American Geographers* 88:352–76.
Peet, R. 1991, *Global Capitalism: Theories of Societal Development*. London: Routledge.
Peet, R. and Watts, M. eds., 1996, *Liberation Ecologies*. London: Routledge.
Peet, R. and Watts, M. 1996 "Liberation Ecology: Development, Sustainability, and Environment in an Age of Market Triumphalism," in R. Peet and M. Watts. eds, *Liberation Ecologies*. London: Routledge.
Pepper, D. 1984, *The Roots of Modern Environmentalism*. London: Croom Helm.
Pepper, D. 1991, *Communes and the Green Vision: Counter-Culture, Lifestyle and the New Age*. London: Green Print.
Pepper, D. 1993, *Eco-Socialism: From Deep Ecology to Social Justice*. London: Routledge.
Pepper, D. 1996, *Modern Environmentalism: An Introduction*. London: Routledge.
Petchesky, R. P. 1987, "Foetal Images: The Power of Visual Culture in the Politics of Reproduction," in M. Stanworth. ed., *Reproductive Technologies*. Cambridge: Polity Press.
Pfeffer, N. 1987, "Artificial Insemination In-Vitro Fertilization and the Stigma of Infertility," in *ibid*.
Phelps, L. 1981, "Patriarchy and Capitalism," in L. Phelps. ed., *Building Feminist Theory: Essays from Quest*. London: Longman.
Phillips, A. 1985, *Divided Loyalties: Dilemmas of Sex and Class*. London: Virago.
Phillips, A. 1991, *Engendering Democracy*. Cambridge: Polity.
Philips, A. 1993, *Democracy and Difference*. Cambridge: Polity.

Phillips, M. 1999, *The Sex-Change Society*. London: Social Market Foundation.
Phizacklea, A. 1990, *Unpacking the Fashion Industry*. London: Routledge.
Pierson, R. R. ed., 1987, *Women and Peace*. Beckenham: Croom Helm.
Plant, J. 1989, *Healing the Wounds: The Promise of Ecofeminism*. Philadelphia, PA: New Society Publishers.
Plant, J. 1997, "The Challenge of Ecofeminist Community," in K. Warren. ed., *Ecofeminism: Women, Nature, Culture*. Philadelphia: New Society.
Plumwood, V. 1988, "Women, Humanity and Nature," *Radical Philosophy* 48:16–24.
Plumwood, V. 1991, *Nature, Self and Gender*. London: Routledge.
Plumwood, V. 1993, *Feminism and the Mastery of Nature*. London: Routledge.
Plumwood, V. 1994, "The Ecopolitics Debate and the Politics of Nature," in K. Warren. ed., *Ecological Feminism*. London: Routledge.
Plumwood, V. 1997, "Androcentrism and Anthropocentrism: Parallels and Politics," in K. Warren. ed., *Ecofeminism: Women, Nature, Culture*. Philadelphia, PA: New Society.
Pollert, A. 1996, "Gender and Class Revisited; or, the Poverty of Patriarchy," *Sociology* 30, 4:639–59.
Porritt, J. 1986, *Seeing Green: The Politics of Ecology Explained*. Oxford: Blackwell.
Porritt, J. and Winner, M. 1988, *The Coming of the Greens*. London: Fontana.
Price, F. 1999, "Beyond Expectation: Clinical Practices and Clinical Concerns," in S. Edwards, J. Edwards, S. Franklin, E. Hirch, F. Price and M. Strathern. eds, 1999 *Technologies of Procreation: Kinship in the Age of Assisted Conception*. Second edition. London: Routledge.
Prigogine, I. 1980, *From Being to Becoming*. San Francisco, CA: Freeman.
Prigogine, I. 1989, "The Philosophy of Instability," *Futures* 21, 4:396–400.
Prigogine, I. and Stengers, I. 1984, *Order Out of Chaos*. New York: Bantam.
Rabinow, P. 1992, "Artificiality and Enlightenment," in J. Crary and S. Kwinter. eds, *Incorporations*. New York: Zone Books.
Ramazanoglu, C. 1989, *Feminism and the Contradictions of Oppression*. London: Routledge.
Ramazanoglu, C. ed., 1993, *Up Against Foucault*. London: Routledge.
Rangan, H. 1996, "From Chipko to Uttaranchal: Development, Environment and Social Protest in the Garhwal Himalayas, India," in R. Peet and M. Watts. eds, *Liberation Ecologies*. London: Routledge.
Raymond, J. 1993, *Women as Wombs: Reproductive Technologies and the Battle over Women's Freedom*. San Francisco, CA: Harper and Row.
Redclift, M. 1984, *Development and the Environmental Crisis: Red or Green Alternatives?* London: Methuen.
Redclift, M. 1987, *Sustainable Development: Exploring the Contradictions*. London: Methuen.
Redclift, M. and Benton, T. eds, 1994, *Social Theory and the Global Environment*. London: Routledge.
Redclift, M. and Woodgate, G. 1994, "Sociology and the Environment: Discordant Discourse?" in M. Redclift and T. Benton. eds, *ibid*.
Reed, E. 1975, *Woman's Evolution: From Matriarchal Clan to Patriarchal Family*. New York: Pathfinder.
Reed, M. and Harvey, D. L. 1992, "The New Science and the Old: Complexity and Realism in the Social Sciences," *Journal for the Theory of Social Behaviour* 22:356–79.

Reed, M. and Harvey, D. L. 1996, "Social Science as the Study of Complex Systems," in L. D. Kiel and E. Elliott. eds, *Chaos Theory in the Social Sciences*. Ann Arbor: University of Michigan Press.

Regan, T. 1988, *The Case for Animal Rights*. London: Routledge.

Rich, A. 1977, *Of Woman Born: Motherhood as Experience and Institution*. London: Virago.

Rich, A. 1980, "Compulsory Heterosexuality and the Lesbian Existence," *Signs* 5, 4: 631–60.

Richards, J. R. 1982, *The Sceptical Feminist*. Harmondsworth: Penguin.

Richardson, D. 1996, "Misguided, Dangerous and Wrong": On the Maligning of Radical Feminism in D. Bell and R. Klein. eds, *Radically Speaking*. London: Zed Books.

Riessman, C. K. 1983, "Women and Medicalization: A New Perspective," *Social Policy* (Summer): 3–18.

Robertson, R. 1992, *Globalization: Social Theory and Global Culture*. London: Sage.

Roiphe, K. 1994, *The Morning After*. London: Hamish Hamilton.

Rollin, B.E. 1981, *Animal Rights and Human Morality*. New York: Prometheus Books.

Rootes, C. ed., 1990, *Environmental Movements: Local, National and Global*. London: Frank Cass.

Rose, H. 1987, "Victorian Values in the Test Tube: The Politics of Reproductive Science and Technology," in M. Stanworth. ed., *Reproductive Technologies*. Cambridge: Polity Press.

Rose, H. 1994, *Love, Power and Knowledge*. Cambridge: Polity.

Rose, H. 2000, "Risk, Trust and Scepticism in the Age of the New Genetics," in U. Beck, 2000 "Risk Society Revisited" in B. Adam, U. Beck, and J. van Loon, J. *The Risk Society and Beyond: Critical Issues for Social Theory*. London: Sage.

Rosenau, J. 1998, "Government and Democracy in a Globalizing World," in D. Archibugi, D. Held and M. Kohler. eds, *Re-Imagining Political Community*. Cambridge: Polity Press.

Roseneil, S. 1995, *Disarming Patriarchy: Feminism and Political Action at Greenham*. Buckingham: Open University Press.

Roseneil, S. 1996, "Transgressions and Transformations: Experience, Consciousness and Identity at Greenham," in N. Charles and F. Hughes-Freeland. eds. *op cit.*

Roszak, T. 1979, *Person/Planet: The Creative Disintegration of Industrial Society*. London: Victor Gollancz.

Roszak, T. 1989, *Where the Wasteland Ends*. Berkeley, CA: Celestial Arts.

Roszak, T. 1992, *The Voice of the Earth: An Exploration in Ecopsychology*. New York: Simon and Schuster.

Rowbotham, S. 1979, "The Trouble with Patriarchy," in M. Evans. ed., *The Woman Question*. London: Fontana.

Rowland, R. and Klein, R. 1996, "Radical Feminism: History, Politics, Action," in D. Bell and R. Klein. eds, *Radically Speaking*. London: Zed Books.

Rubin, G. 1975, "The Traffic in Women: Notes on the 'Political Economy' of Sex," in R. Reiter. ed., *Toward An Anthropology of Women*. New York: Monthly Review Press.

Ruddick, S. 1990, *Maternal Thinking*. London: Women's Press.

Rudy, A. P. 1998, "Ecology and Anthropology in the Work of Murray Bookchin: Problems of Theory and Evidence," in A. Light. ed., 1998, *Social Ecology After Bookchin*. New York: The Guilford Press.
Russell, D. 1982, *Rape in Marriage*. New York: Macmillan.
Russell, D., Radford, J. 1994, *Femicide: The Politics of Woman Killing*. London: Hutchinson.
Ryan, A. 1970, *The Philosophy of the Social Sciences*. London: Macmillan.
Ryle, M. 1988, *Ecology and Socialism*. London: Century Hutchinson.
Sachs, W. 1993, ed., *Global Ecology*. London: Zed Books.
Said, E. 1978, *Orientalism*. New York: Pantheon.
Said, E. 1983, *The World, the Text, and the Critic*. London: Vintage.
Said, E. 1988, "Michel Foucault 1926–1984" in J. Arac. ed., *op cit*.
Salamone, C. 1982, "The Prevalence of the Natural Law Within Women: Women and Animal Rights," in P. McAllister. ed., *Reweaving the Web of Life*. San Francisco: New Society Publishers.
Sale, K. 1980, *Human Scale*. New York: Coward, Cann and Geoghegan.
Sale, K. 1985, *Dwellers in the Land: The Bioregional Vision*. San Francisco, CA: Sierra Club Books.
Salleh, A. K. 1984, "Deeper than Deep Ecology: The Eco-Feminist Connection," *Environmental Ethics* 6, 4:339–45.
Salleh, A. K. 1997, *Ecofeminism as Politics: Nature, Marx and the Postmodern*. London: Zed.
Salomonsen, J. 2002, *Enchanted Feminism: The Reclaiming Witches of San Francisco*. London: Routledge.
Sargent, L. ed., 1981, *Women and Revolution*. London: Pluto.
Sawicki, J. 1988, "Feminism and the Power of Foucauldian Discourse," in J. Arac. ed., *op cit*.
Sawicki, J. 1991, *Disciplining Foucault: Feminism, Power, and the Body*. London: Routledge.
Sayer, A. 1992, *Method in Social Science: A Realist Approach*. London: Routledge.
Sayer, A. 2000, *Realism and Social Science*. London: Sage.
Sayers, J. 1982, *Biological Politics: Feminist and Anti-Feminist Perspectives*. London: Tavistock.
Scarry, E. 1985, *The Body in Pain: The Making and Unmaking of the World*. Oxford: Oxford University Press.
Schnailberg, A. 1980, *The Environment, from Surplus to Scarcity*. Oxford: Oxford University Press.
Schnailberg, A. and Gould, K. 1994, *Environment and Society: The Enduring Conflict*. New York: St. Martins Press.
Schumacher, E. 1976, *Small is Beautiful*. London: Sphere.
Schwichentenberg, C. 1993, *The Madonna Complex: Representational Politics, Subcultural Identities, and Cultural Theory*. Oxford: Westview Press.
Scott, A. 2000, "Risk Society or Angst Society?" in U. Beck. 2000, "Risk Society Revisited" in B. Adam, U. Beck, and J. van Loon, J. *The Risk Society and Beyond: Critical Issues for Social Theory*. London: Sage.
Seabrooke, J. 1986, "The Inner City Environment: Making the Connections," in J. Weston. ed., *Red and Green*. London: Pluto.
Segal, L. 1987, *Is the Future Female? Troubled Thoughts on Contemporary Feminism*. London: Virago.

Segal, L. 1990, *Slow Motion: Changing Masculinities, Changing Men*. London: Virago.
Segal, L. and McIntosh, M. eds, 1992, *Sex Exposed: Sexuality and the Pornography Debate*. London: Virago.
Segal, L. 1994, *Straight Sex: The Politics of Pleasure*. London: Virago.
Segal, L. ed., 1997, *New Sexual Agendas*. London: Macmillan.
Segal, L. 1997, "Feminist Sexual Politics and the Heterosexual Predicament," in L. Segal. ed., *New Sexual Agendas*. London: Macmillan.
Sen, G. and Grown, C. 1987, *Development, Crises and Alternative Visions*. New York: Monthly Review Press.
Sessions, G. ed., 1995, *Deep Ecology for the Twenty First Century*. Boston: Shambhala Press.
Sharpe, K. 1992, "Biology and Social Science: A Reply to Ted Benton," *Sociology* 26, 2: 219–25.
Shaviro, S. 1995, "Two Lessons from Burroughs," in J. Halberstram and I. Livingston. eds, *Posthuman Bodies*. Bloomington, IL: Indiana University Press.
Sheridan, A. 1980, *Michel Foucault: The Will to Truth*. London: Tavistock.
Shilling, C. 1993, *The Body and Social Theory*. London: Sage.
Shilling, C. 2001, "Embodiment, Experience and Theory: In Defence of the Sociological Tradition," *The Sociological Review* 49, 3:327–44.
Shilling, C. 2003, *The Body and Social Theory*. Second edition. London: Sage.
Shiva, V. 1988, *Staying Alive: Women, Ecology and Development*. London: Zed Press.
Shiva, V. 1993, *Monocultures of the Mind*. London: Zed Books.
Shiva, V. 1998, *Biopiracy: The Plunder of Nature and Knowledge*. Dartington: Green Books.
Shiva, V. 2000, "The World on the Edge," in W. Hutton and A. Giddens. eds, *Global Capitalism*. New York: New Press.
Short, J. R. 1991, *Imagined Country: Environment, Culture and Society*. London: Routledge.
Shuttle, P. and Redgrove, P. 1986, *The Wise Wound: The Myths, Realities, and Meanings of Menstruation*. New York: Grove Press.
Singer, P. 1979, *Practical Ethics*. Boston: Cambridge University Press.
Singer, P. 1981, *The Expanding Circle: Ethics and Sociobiology*. New York: Farrar, Strauss and Giroux.
Singer, P. ed., 1985, *In Defence of Animals*. Oxford: Blackwell.
Singer, P. 1990, *Animal Liberation*. Fourth edition. New York: Avon Books.
Skeggs, B. 1997, *Formations of Class and Gender: Becoming Respectable*. London: Sage.
Sklair, L. 1991, *Sociology of the Global System*. Hemel Hempstead: Harvester Wheatsheaf.
Sklair, L. 1994, "Global Sociology and Global Environmental Change," in M. Redclift and T. Benton. eds, *Social Theory and Global Environmental Change*. London: Routledge.
Slack, J. D. and Whitt, A. 1994, "Communities, Environments and Cultural Studies," *Cultural Studies* 8, 1:5.
Slater, D. 1992, "On the Borders of Social Theory," *Society and Space* 10: 307–27.
Slicer, D. 1994, "Wrongs of Passage: Three Challenges to the Maturing of Ecofeminism," in K. J. Warren. ed., *Ecological Feminism*. London: Routledge.

Smart, C. 1984, *The Ties that Bind: Law, Marriage and the Reproduction of Patriarchal Relations*. London: Routledge and Kegan Paul.

Smart, C. 1987, "There is of course the Distinction Dictated by Nature': Law and the Problem of Paternity," in M. Stanworth. ed., *Reproductive Technologies*. Cambridge: Polity.

Smith, A. 1997, "Ecofeminism Through an Anticolonial Framework," in K. Warren. ed., *Ecofeminism: Women, Nature, Culture*. Philadelphia, PA: New Society.

Smith, E. 1995, "Crossing the Great Divide: Race, Class and Gender in Southern Women's Organizing, 1979–1991," *Gender and Society* 9:6.

Smith, P. 1988, *Discerning the Subject*. Minneapolis: University of Minnesota Press.

Sobchack, V. 1995, "Beating the Meat/Surviving the Text, or How to Get Out of the Century Alive," in M. Featherstone and R. Burrows. eds, *Cyberspace/Cyberbodies/Cyberpunk: Cultures of Technological Embodiment*. London: Sage.

Sokal, A. and Bricmont, J. 1998, *Intellectual Impostures*. London: Routledge.

Solonas, V. 1983, *SCUM Manifesto*. San Francisco, CA: Phoenix Press, first published 1968.

Sontheimer, S. ed., 1991, *Women and the Environment: Crisis and Development in the Third World*. London: Earthscan.

Soper, K. 1989, "Feminism as Critique," *New Left Review* 176:91–114.

Soper, K. 1990, "Postmodernism, Subjectivity and the Question of Value," *New Left Review*, 186:120–28.

Soper, K. 1993, "Productive Contradictions," in C. Ramazanoglu. ed., *Up Against Foucault*. London: Routledge.

Soper, K. 1994, "Feminism, Humanism and Postmodernism," in M. Evans. ed., *The Woman Question*. Second edition. London: Sage.

Soper, K. 1995, *What is Nature?*. Oxford: Blackwell.

Soper, K. 1996a, "Greening the Prometheus: Marxism and Ecology" in T. Benton. ed., *The Greening of Marxism*. London: Guilford Press.

Soper, K. 1996b, "Feminism, Ecosocialism and the Conceptualization of Nature" in T. Benton. ed., *The Greening of Marxism*. London: Guilford.

Spallone, P. 1987, *Made to Order: The Myth of Reproductive and Genetic Progress*. New York: Pergamon.

Spelman, E.V. 1990, *Inessential Woman: Problems of Exclusion in Feminist Thought*. London: The Women's Press.

Spender, D. 1980, *Man Made Language*. London: Routledge.

Spender, D. 1983, *Women of Ideas*. London: Ark.

Spender, D. 1985, *For the Record: The Making and Meaning of Feminist Knowledge*. London: Women's Press.

Spiegal, M. 1988, *The Dreaded Comparison: Human and Animal Slavery*. London: Routledge.

Spivak, G. C. 1987, *In Other Worlds: Essays in Cultural Politics*. London: Routledge.

Spretnak, C. ed., 1982, *The Politics of Women's Spirituality: Essays on the Rise of Spiritual Power within the Women's Movement*. New York: Doubleday/Anchor.

Spretnak, C. 1985, "The Spiritual Dimension of Green Politics," in C. Spretnak and F. Capra. eds, *Green Politics*. Glasgow: Paladin.

Spretnak, C. and Capra, F. 1985, *Green Politics: The Global Promise*. Glasgow: Paladin.

Spretnak, C. 1989, "Towards an Ecofeminist Spirituality," in J. Plant. ed., *Healing the Wounds: The Promise of Ecofeminism*. Philadelphia, PA: New Society Publishers.
Spretnak, C. 1990, "Ecofeminism: Our Roots and Flowering," in I. Diamond, and G. Orenstein. eds, *Reweaving the World*. San Francisco, CA: Sierra Club Books.
Spretnak, C. 1991, *States of Grace*. New York: HarperCollins.
Spretnak, C. 1996, "The Disembodied Worldview of Deconstructive Postmodernism," in D. Bell and R. Klein. eds, *Radically Speaking*. London: Zed Books.
Spretnak, C. "Radical Non-Duality in Ecofeminist Philosophy," in K. Warren. ed., *Ecofeminism: Women, Nature, Culture*. Philadelphia, PA: New Society.
Stabile, C. A. 1994a, " 'A Garden Enclosed is My Sister': Ecofeminism ad Eco-Valences," *Cultural Studies* 8, 1:56–73.
Stabile, C. A. 1994b, *Feminism and the Technological Fix*. Manchester: Manchester University Press.
Stacey, J. 1997, *Terratologies: A Cultural Study of Cancer*. London: Routledge.
Stanko, E. 1985, *Intimate Intrusions: Women's Experience of Male Violence*. London: Routledge.
Stanley, L. 1990, *Feminist Praxis: Research, Theory and Epistemology in Feminist Sociology*. London: Routledge.
Stanworth, M. 1987, *Reproductive Technologies*. Cambridge: Polity.
Stanworth, M. 1987, "Reproductive Technologies and the Deconstruction of Motherhood," in *ibid*.
Starhawk, 1989, *The Spiral Dance: A Rebirth of the Ancient Religion of the Great Goddess*. New York: Harper & Row first published in the USA 1979.
Starhawk, 1990a, *Dreaming the Dark: Magic, Sex and Politics*. London: Unwin, first published in the USA 1982.
Starhawk, 1990b, *Truth or Dare*. San Francisco: HarperCollins.
Stone, M. 1977, *The Paradise Papers: The Suppression of Women's Rites*. London: Virago.
Strathern, M. 1988, *The Gender of the Gift*. Berkeley, CA: University of California Press.
Strathern, M. 1992, *Reproducing the Future: Essays on Anthropology, Kinship and the New Reproductive Technologies*. Manchester: Manchester University Press.
Strathern, M. 1993, "A Question of Context," in J. Edwards, S. Franklin, E. Hirsch, F. Price and M. Strathern. eds, *Technologies of Procreation*. London: Macmillan.
Strathern, M. 1999, "Regulation, Substitution and Possibility," in J. Edwards, S. Franklin, E. Hirsch, F. Price and M. Strathern. eds, *Technologies of Procreation: Kinship in the Age of Assisted Conception*. Second edition. London: Routledge.
Sturgeon, N. 1997a, "The Nature of Race: Discourses of Racial Difference in Ecofeminism," in K. Warren. ed., *Ecofeminism: Women, Nature, Culture*. Philadelphia: New Society.
Sturgeon, N. 1997b, *Ecofeminist Natures: Race, Gender, Feminist Theory and Political Action*. London: Routledge.
Taylor, D. E. 1997, "Women of Color, Environmental Justice, and Ecofeminism," in K. Warren. ed., *Ecofeminism: Women, Nature, Culture*. Philadelphia, PA: New Society.
Tester, K. 1991, *Animals and Society: The Humanity of Animal Rights*. London: Routledge.

Thomas, K. 1971, *Religion and the Decline of Magic*. London: Weidenfield & Nicholson.
Thomas, K. 1983, *Man and the Natural World: Changing Attitudes in England. 1500–1800* London: Allen Lane.
Thompson, D. 1996, "The Self-Contradiction of 'Post-Modernist' Feminism" in D. Bell and R. Klein. eds, *Radically Speaking*. London: Zed Books.
Thompson, J. and Held, D. eds, 1982, *Habermas: Critical Debates*. London: Macmillan.
Tobias, M. ed., 1985, *Deep Ecology*. San Diego: Avant.
Tokar, B. 1987, *The Green Alternative*. San Pedro, CA: R. and E. Miles.
Tokar, B. 1988, "Social Ecology, Deep Ecology and the Future of Green Thought," *The Ecologist* 18: 4–5.
Tong, R. 1989, *Feminist Thought*. London: Unwin Hyman.
Trainer, T. 1985, *Abandon Affluence!* London: Zed Books.
Turner, B. 1984, *The Body and Society*. London: Sage.
Turner, B. 1991, "Recent Developments in the Theory of the Body," in M. Featherstone, M. Hepworth and B. Turner. eds, *The Body: Social Process and Cultural Theory*. London: Sage.
Turner, B. 1992, *Regulating Bodies: Essays in Medical Sociology*. London: Routledge.
Urry, J. 1995, *Consuming Places*. London: Routledge.
Urry, J. 2000, *Sociology Beyond Societies: Mobilities for the Twenty-First Century*. London: Routledge.
Ussher, J. M. 1997, "The Case of the Lesbian Phallus: Bridging the Gap between Material and Discursive Analyses of Sexuality," in L. Segal. ed., *New Sexual Agendas*. London: Macmillan.
Vance, C. S. ed., 1984, *Pleasure & Danger: Exploring Female Sexuality*. London: Routledge and Kegan Paul.
Vance, C. S. 1984, "Pleasure and Danger: Towards a Politics of Sexuality," in C. S. Vance. ed., *ibid.*
Vance, C. 1989, "Social Construction Theory: Problems in the History of Sexuality," in A. van Kooten Nierker and T. van de Meer. eds, *Homosexuality, which Homosexuality?* London: GMP Publishers.
Vance, L. 1995, "Beyond Just-So Stories: Narrative, Animals and Ethics," in C. J. Adams and J. Donovan. eds. *op cit.*
Varela, F. Thompson, E. and Rosch, E. 1991, *The Embodied Mind: Cognitive Science and Human Experience*. Cambridge, Mass.: MIT Press.
Vergata, A. 1994, "Herbert Spencer: Biology, Sociology, and Cosmic Evolution," in S. Maasan. ed., *Biology as Society, Society as Biology: Metaphors*. Dordrecht: Kluwer Academic Press.
Vogel, L. 1983, *Marxism and the Oppression of Women*. London: Pluto Press.
Walby, S. 1986, *Patriarchy at Work*. Cambridge: Polity.
Walby, S. 1988, "Gender Politics and Social Theory," *Sociology* 22,2:215–32.
Walby, S. 1990, *Theorizing Patriarchy*. Oxford: Basil Blackwell.
Walby, S. 1992, "Post-Post-Modernism? Theorizing Social Complexity," in M. Barrett and A. Phillips. eds, *Destabilizing Theory: Contemporary Feminist Debates*. Cambridge: Polity Press.
Walby, S. 1997, *Gender Transformations*. London: Routledge.
Walby, S. 1999, "The New Regulatory State: The Social Powers of the European Union," *British Journal of Sociology*, 50,1:118–40.

Walby, S. 2003a, "Complexity Theory, Globalization and Diversity," paper presented to conference of the British Sociological Association, University of York, April 2003.
Walby, S. 2003b, "Modernities/Globalization/Complexities," paper presented to conference of the British Sociological Association, University of York, April 2003.
Walby, S. 2003c, "The Myth of the Nation State: Theorizing Societies and Polities in the Global era," *Sociology* 37, 3:531–48.
Walby, S. 2004, "The European Union and Gender Equality: Emergent Varieties of Gender Regime," *Social Politics* 11, 1:4–29.
Wall, D. 1999, *Earth First! and the Anti-Roads Movement*. London: Routledge.
Wallerstein, I. 1974, *The Modern World System: Capitalist Agriculture and the Origins of the European World Economy in the Sixteenth Century*. London: Academic Press.
Wallerstein, I. 1979, *The Capitalist World Economy*. Cambridge: Cambridge University Press.
Wallerstein, I. 1990, "Societal Development, or Development of the World System?" in M. Albrow and E. King. eds, *Globalization, Knowledge and Society* London: Sage.
Walter, N. 1998, *The New Feminism*. London: Little Brown.
Warren, K. 1987, "Feminism and Ecology: Making Connections," *Environmental Ethics* 9:3–20.
Warren, K. 1990, "The Power and the Promise of Ecological Feminism," *Environmental Ethics* 12:125–46.
Warren, K. (guest ed.,) 1991, *Hypatia: Special Issue on Ecological Feminism*. 6,1.
Warren, K. 1993, "A Philosophical Perspective on Ecofeminist Spiritualities," in C. J. Adams. ed., *Ecofeminism and the Sacred*. New York: Continuum Books.
Warren, K. 1994, *Ecological Feminism*. London: Routledge.
Warren, K. 1997, "Taking Empirical Data Seriously: An Ecofeminist Philosophical Perspective" in Warren. ed., *Ecofeminism: Women, Nature, Culture*. Philadelphia, PA: New Society.
Warren, K. 2000, *Ecofeminist Philosophy*. Savage, MD: Rowland and Littlefield/ Macmillan.
Waters, K. 1996, "Returning to the Modern: Radical Feminism and the Postmodern Turn," in D. Bell and R. Klein. eds, *Radically Speaking*. London: Zed Books.
Watson, D. 1998, "Social Ecology and the Problem of Technology," in A. Light. ed., 1998, *Social Ecology After Bookchin*. New York: The Guilford Press.
Weedon, C. 1987, *Feminist Practice and Poststructuralist Theory*. Oxford: Basil Blackwell.
Weeks, J. 1981, *Sex, Politics and Society*. London: Longman.
Weeks, J. 1989, *Sexuality and its Discontents*. London: Routledge.
Weitz, R. 2003, "A History of Women's Bodies," in R. Weitz. ed., *The Politics of Women's Bodies: Sexuality, Appearance and Behavior*. Second edition. Oxford: Oxford University Press.
Weitz, R. ed., *The Politics of Women's Bodies: Sexuality, Appearance and Behavior*. Second edition. Oxford: Oxford University Press.
Weitz, R. 2001, "Women and Their Hair: Seeking Power Through Resistance and Accommodation," *Gender and Society* 15, 5:667–86.
Weston, J. ed., 1986, *Red and Green: A New Politics of the Environment*. London: Pluto.

Whatmore, S. 1997, "Dissecting the Autonomous Self," *Environment and Planning D: Society and Space* 15:37–53.

Whatmore, S. 1999, "Hybrid Geographies: Rethinking the 'Human' in Human Geography" in D. Massey, J. Allen and P. Sarre. eds, *Human Geography Today*. Cambridge: Polity.

Whelehan, I. 2000, *Overloaded: Popular Culture and the Future of Feminism*. London: The Women's Press.

Williams, L. 1990, *Hard Core: Power, Pleasure and the Frenzy of the Visible*. London: Pandora.

Williams, L. 1992, "Pornographies on/Scene, or Diff'rent Strokes for Diff'rent Folks," in L. Segal and M. McIntosh. eds, *Sex Exposed*. London: Virago.

Wilson, E. 1992, "Feminist Fundamentalism: The Shifting Politics of Sex and Censorship," in L. Segal and M. McIntosh *ibid*.

Wilson, E. O. 1984, *Biophilia*. Cambridge, MA: Harvard University Press.

Wilson, E. O. and Peter F. M. eds, 1988, *Biodiversity*. Washington DC: National Academy Press.

Wilton, T. 1996, "Genital Identities: An Idiosyncratic Foray into the Gendering of Sexualities," in L. Adkins and V. Merchant. eds, *Sexualizing the Social: Power and the Organization of Sexuality*. Explorations in Sociology 47. London: Macmillan.

Wolf, N. 1990, *The Beauty Myth*. London: Vintage.

Wolf, N. 1993, *Fire with Fire*. London: Chatto and Windus.

Wynne, B. 1992, "Risk and Social Learning: Reification to Engagement," in S. Krimsky, and D. Golding. eds, *Social Theories of Risk*. Westport, CT.: Praeger.

Wynne, B. 1994, "Scientific Knowledge and the Global Environment," in M. Redclift and T. Benton. eds, *op cit*.

Wynne, B. 1996, "May the Sheep Safely Graze? A Reflexive View of the Expert-Lay Knowledge Divide," in S. Lash, B. Szerszynski and B. Wynne. eds, *op cit*.

Yearley, S. 1992, *The Green Case: A Sociology of Environmental Issues, Arguments and Politics*. London: Routledge.

Yearley, S. 1996, *Sociology, Environmentalism, Globalization*. London: Sage.

Young, I. 1981, "Beyond the Unhappy Marriage: A Critique of Dual Systems Theory," in L. Sargent. ed., *Women and Revolution: The Unhappy Marriage of Marxism and Feminism*. London: Pluto.

Young, I. 1990a, "The Ideal of Community and the Politics of Difference," in L. Nicholson. ed., *Feminism/Postmodernism*, first published 1987. London: Routledge.

Young, I. 1990b, *Justice and the Politics of Difference*. Oxford: Princeton, NJ University Press.

Young, I. 1990c, *Throwing Like a Girl*. Bloomington: Indian University Press.

Young, I. 2003, "Breasted Experience: The Look and the Feeling," in R. Weitz. ed., *The Politics of Women's Bodies: Sexuality, Appearance and Behavior*. Second edition. Oxford: Oxford University Press.

Young, J. 1989, *Postenvironmentalism*. London: Belhaven Press.

Zipper, J., Sevenhuijsen, S. 1987, "Surrogacy: Feminist Notions of Motherhood Reconsidered," in M. Stanworth. ed., *Reproductive Technologies*. Cambridge: Polity.

Zita, J. 1996, "The Male Lesbian and the Post-Modernist Body," in C. Card. ed., *Hypatia: A Journal of Feminist Philosophy* Special Issue: Lesbian Philosophy 7, 4:106–27.

Index

actor-network theory 53
Adams, Carol 147–8
affinity ecofeminism 102, 104
agriculture 140
androcentrism 122
animal bodies 135
animal rights 48, 106
animals 149–50
  alienation 140
  assisted conception 144–5
  violence against 173
animistic societies 110
anthroparchy 8, 14, 45, 54, 63–5, 159, 161, 164, 166–7, 170, 171, 175
  cultures of exclusive humanism 69–70, 169–70
  politics 68–9, 168
  production relations of 65–7, 167
  reproduction and domestication 67–8, 167–8, 174–5
anthropocentrism 8, 13, 17–19, 40, 64, 122, 166
  and Marxism 32
autopoiesis 56, 59–60, 90, 154, 161

Beck, Ulrich 48–9
Benton, Ted 31, 33
biocentric egalitarianism 20
biological determinism 136
biotechnology 32, 38, 67, 139, 140, 147, 155
black feminism 76–7, 79, 83, 95–6, 162–3
body, physical
  gendered 131–4
  marginalization 136
  postmodern theories 83
  poststructuralist theories 130–1
body, symbolic 131
Bookchin, Murray 24–5, 33
  evolutionary hypothesis 26–8
  social ecology 28–9, 40
  social hierarchy 28

Bourdieu, Pierre 75
Braidotti, Rosi 149–50
Butler, Judith 84–5, 91
Byrne, David 44, 62

capitalism 30–2, 161, 167
  alienation under 139
  global domination of 66
  and oppression of women 72, 73–6, 115
  and patriarchy 123–4
Capra, Fritjof 51, 58, 60, 62, 63, 67
Castells, Manuel 66
Castree, Noel 31, 32, 46
chaos 4–5, 39–40, 43–4, 137, 176
class 75, 76
co-constructionism 52–3
coevolution 27, 153
Collard, Andrée 112
Collins, Patricia Hill 76–7
complexity theory 4–5, 14, 16, 28, 41, 43–4, 49, 70, 154, 156, 159, 163–4, 176
  living systems 54–8
  ontological and epistemological implications 44
  and patriarchy 86–90
  social systems 59–63
conception, assisted 143–5
conservation imperialism 35, 37
corporeality 3, 6, 83, 85, 128, 129, 149
critical realism 3–4, 43–4, 46, 50–3, 60–1, 164–5
cultural imperialism 74
cultural institutions
  role in male dominance 80
culture 170–1
  meat eating cultures 148
  patriarchal 97
cyborgs 39, 47, 82, 114, 130, 145–7, 148, 151, 152–3

213

Daly, Mary 104, 107, 108, 112, 113, 116
De Beauvoir, Simone 102
de-development 22–3, 37
deep ecology 5, 6, 8, 17–18, 29, 34, 37, 40, 44, 69, 101, 153
 and differences 20–1
 and embodied humans 135–6
 and social differences 21–3
Devall, Bill 16, 19, 20, 21, 22, 136
Dickens, Peter 51, 52, 139–40
difference feminists *see* feminism, postmodern
differences 5–7, 14, 16, 25, 83–4, 126–7, 157
 and deep ecology 20–1
 Luhman's concept of 90–1
differences-in-domination 25, 29, 86, 141, 155
 multiplicity of 93
differential value 21
discourses 104, 132, 158–9, 160, 162
 and domination 104–5
 on femininity 97
discursive co-constitution 135, 147
dissipative structures 55, 65, 176–7
diversity 20–1, 29, 176–8
domesticity 113
 gendered 171
domination
 hierarchical system of 24–6, 28–9
 three levels of 7
dualistic conceptual frameworks 121–2

ecofeminism 1–2, 14–15, 33, 41, 45, 72, 99–100, 101–2, 165
 critiques 102–3, 112–14, 119
 and patriarchy 103
 and pre-historical anthropology 109–11
ecological self 20
ecologisms 13, 16–17, 157
 Bookchin's 24–6
 kinds 16
 versus environmentalism 16
eco-socialism 17, 30, 66
 impact of Marxism 30–2
ecosophy 17

embodied materialism 3, 128–30, 154–5
embryo transfer 138, 143, 144
employment 80, 81, 96–7, 174
Engels, Friedrich 139
environment 12–13, 164, 171
 conceptualization 45
environmental degradation 53–4, 66
 feminist theory 101–2
 and mistreatment of women 3
 and social inequality 33, 34
environmental ethics 19
 hierarchical approach 26, 28
environmental exploitation 32, 34, 122–3
 impact of mechanistic science 117
environmentalism 16
 of the poor 36–8
environmental justice 34, 177
 Third world 36
environmental protests 33
environmental risks 48–9
environment-society relations 1, 2, 26, 29
 and gender 33
 impact of social differences on 16
 regional discursive formations 35–6
Escobar, Arturo 38–9
essentialism 4, 72, 82–3
 ecofeminist 103, 112–14, 119–20
Eurocentrism 122
evolution 26–7
 Gaian theory 18, 57–8
exploitation 7, 17, 33, 64, 74

femininity
 discourses 97, 105
feminisms 13–14, 72–3, 80–1
 postmodern 5, 6, 71, 81–2, 99
filiarchy 125
Foucault, Michel 104–5, 130, 131, 132, 134
Fox, Warwick 19, 120
Friends of the Earth 36

Gaia evolutionary theory 18, 57–8
Gandhi, Mohandas K. 34

gender 13
  and class 75, 76
  conceptualization of 9–10, 11, 82–3
  cultural formation of 15
  and environment-society relations 33
  patriarchal discourses 105–8
  as performance 84–6
  and race 76–7
  systemic and structural approaches 72–3, 114–16
gender domination 8–9, 14, 43
gendered structures 87
  transformative quality 88–9
gender hierarchy 72
gender inequality 74
gender regimes 89
  see also patriarchy
gender relations 100
  patriarchal system of 77–81, 99, 100, 165–6
genetic engineering 67, 139, 140, 146
geophysics 56–7
Giddens, Anthony 61, 85
globalization 66, 87–8, 147
goddess worship 109, 110
Greer, Germaine 80–1
Griffin, Susan 104, 107, 108, 112–13
group oppression 74
Grosz, Elizabeth 84, 131
Guha, Ramachandra 34, 35, 36, 37

Halberstram, Judith 150
Haraway, Donna 39, 117–18, 130, 145–7, 148, 149, 151, 152–3, 154
Hayles, Katherine N. 4, 44, 134, 146, 151, 152, 156
heterosexuality 79, 85, 172
hierarchical thinking 120
hierarchy 26–7
Hinduism 109
human beings 42–3
  relationship with animals 21, 106, 134–7, 153
human domination 8, 65, 167
  see also anthropocentrism
  over nature 15, 17, 19, 23–4, 26
  social system of 43

human rights 33, 36, 43, 58, 88, 89, 177
hybridity 15, 39, 52, 53, 129, 138, 145, 146, 147, 151, 152–3, 165
  see also cyborgs
hypercomplexity 93, 158

identity 75, 91
  and differences 84
interpenetration 3, 12, 65, 69, 92, 93–4, 117, 147, 158, 165, 167, 175
intra-human domination 16, 18, 23, 24
  and anthropocentrism 19
  and deep ecology 21, 23
intrinsic value 19–20, 28
in vitro fertilization (IVF) 142, 143, 144, 145

Kaufman, Stuart 28
Kelly, Petra 98, 101
  see also material and/or materialism
Kropotkin, Peter 26, 27

labor 139
land ethic 21
liberation ecologism 17, 34–5, 37, 44–5
  postmodern analytics 38–40
living systems
  complexity in 54–8
Lovelock, James 18, 56–7, 58
Luhmann, Niklas 59–60, 62, 86, 90–2, 93, 158

male domination 8, 107
  deconstruction of 81–3
  role of cultural institutions 80
male violence 79, 98
marginalization 7, 64, 71, 74, 111
Margulis, Lynn 56, 57
Martinez-Alier, Juan 34, 37
Marxism 30–2
Marxist feminism 73–4
Marx, Karl 56, 139
masculinity 105
material and/or materialism 3, 11, 12, 13, 15, 23, 31, 35, 38, 49, 52, 60, 62, 68, 70

material semiotic actors 152–3
matriliny 78, 110
matrix 86, 93, 163, 177, 178
Maturana, Humberto 56
Mellor, Mary 124–6
Merchant, Carolyn 112, 114, 115, 116–17, 141
Mies, Maria 122–4
mixed communities 21, 23, 136, 151
modernity, Western
  impact on underdeveloped countries 115–16
  violence against animals 173
multiple systems of social domination 2, 3, 9, 15, 29, 94, 100, 127, 157, 162–4
multiple systems theory 2, 8, 9, 15, 89

Naess, Arne 17, 18, 19, 20, 21, 136, 151
nation state 173–4, 177
  and patriarchy 98–9
natural ecosystems 18
natural structures 50–1, 161
  complexity in 54–8
nature 11–12, 13, 15, 38
  colonization of 35–6
  conceptualization 42, 45
  human domination over 15, 17, 18, 23–4, 26
  industrializing 140
  patriarchal discourses 105–8
  postmodern approaches 46–7
  realism and 50–3, 60–1
  soft constructionism 48–9
  social constructionism 4, 29, 33, 43, 45–6, 50, 60
  social domination of 63–5
  systemic and structural approaches 114–16
  and women 101–2, 106–8, 126
neo-liberalism 66
neo-pagan ecofeminism 110–12
nongovernmental organizations (NGOs) 36–7

open systems *see* dissipative structures
oppression 7, 8, 33, 64, 83–4, 108, 121, 165

faces of 74
multiplicity of 76–7, 86
systems of 160, 162
of women 9, 24, 73–6
paganism 109
Parsons, Talcott 55, 56
patriarchal discourses
  of gender and nature 105–8, 115
patriarchal structures 80–1, 88, 94–5, 170, 175
  culture 97, 170–1
  domestication 174–5
  household 95–6, 174
  paid employment 96–7, 174
  sexuality 95, 171–2
  the state 98–9, 173–4
  violence 97–8, 172–3
patriarchy 8–9, 14, 25, 71–2, 73, 77–8, 88, 94, 157, 162, 164, 166, 170–1
  and capitalism 123–4
  and complexities of domination 86–7
  complex theories 87–90, 98–9
  conceptual framework 120–1
  ecofeminist view 103
  origins 78
  and postcolonialism 115–16
  postmodernist critiques 81–2, 83, 100
  theories 78–9
Pepper, David 31, 33
Plumwood, Val 121–2
politics, anthroparchal 68–9, 168
Pollert, Anna 74, 98
popular culture
  role in promotion of male dominance 80, 97
  sexuality in 95
postcolonialism 35–6, 38–9, 68, 115–16, 161
posthumanism 15, 130, 151–4
  towards 148–51
postmodernism 5, 14, 71, 81–2, 99, 119
  and patriarchy 81–2, 83, 100
poststructuralism 6, 35, 38
poverty 30, 34, 37, 38, 44, 80
power relations 120, 160

Prigogine, Ilya 28, 55, 56
privilege 76–7, 120
production
 anthroparchal relations in 65–7, 167
 gendered 138–41
 patriarchal relations of 96

race and gender 76–7
radical feminism 8, 77–81, 102, 103
 and procreation 142
religious symbolism 110–11
reproduction (biological)
 alienated 137–8, 139
 anthroparchal 67–8, 167–8
 Western model 142
reproductive technologies 139, 141–2, 172
 gender and nature 142–5
Roszack, Theodore 33

Sagan, Dorion 57
Salleh, Ariel 2, 3, 23, 33, 112
scientific knowledge
 impact on gender relations and nature 114–15, 116–19
sentiency 21, 28, 29, 69
sex/gender distinction 9–10
sexuality 80
 and patriarchy 95, 171–2
Singer, Peter 42–3
Skeggs, Beverly 75
social Darwinism 26
social differences 14, 41
 and deep ecology 21–3
 impact on environment-society relations 16
social domination 1, 41
 of gender 86
 interconnections between different forms of 119–27
 multiplicity 2, 3, 9, 15, 29, 43, 94, 100, 157, 162–4
 of nature 19, 63–5
 systems of 24–6, 161–2
social ecofeminism 102
social ecology 17, 23–4, 40–1, 44
 Bookchin's 28–9, 40
social hierarchies 24
 emergence of 24–5, 27, 28

social inequality 32–3, 34
socialist feminism 73
social structures 50–1, 86–7, 91–2, 123, 164
 expectational paradigm 92
social systems 86, 157, 158, 160, 161–2, 164, 176
 complexity in 59–63, 90–3
society 42
species being 31, 140
Spinoza, Benedict de 18
structural analyses 4–5
structural coupling 2, 92, 154
structuration theory 61, 85–6
structures 158–60, 162
sustainable development 37, 38, 39
symbiogenesis 57
symbolic regimes of domination 3
systemic analyses 4–5, 71
systems 157–8
systems thinking 14, 44, 54

terraforming 140–1
Tester, Keith 47–8
Turner, Bryan 134–5

United Nations (UN) 177
urban dwellers 22

value-dualisms 120
Vandana Shiva 67–8, 114, 115–16, 117, 141, 147, 148
violence 68–9, 74, 160, 168–9, 172–3
 against women 79, 81
 patriarchal 97–8, 172

Walby, Syliva 71, 87, 88, 96, 98, 99, 133
Warren, Karen 120–1
wilderness 22, 23, 34, 64, 128
witchcraft, European 110
Woman and Nature (Griffin) 113
women 10–11, 13
 control of sexuality and reproduction of 138
 employment 80, 81, 96–7, 123
 empowerment 112–13

women – *continued*
  exploitation 122–3
  imposed altruism 124–5
  and nature 101–2, 106–8, 126
  violence against 79, 172
women's oppression 9, 24
  and capitalism 72, 73–6, 115
  and scientific revolution 114–15
  and social structures 80–1
World Trade Organization (WTO) 66, 67, 177

Young, Iris Marion 74